THE ROUGH GUIDE to
Apple Watch

The essential guide to the ultimate smartwatch

by Dwight Spivey

Penguin Random House

Book editor: Rachel Mills

Acquisitions editor: Janette Lynn

Book designer: William Thomas

Development editor: Ann Barton

Cover designer: Richard Czapnik

Additional design input: Nikhil Agarwal, Richard Czapnik, Dan May

Photographer: Ralph Anderson

Proofreader/Angliciser: Liz Jones

Indexer: Johnna VanHoose Dinse

Managing editor: Mani Ramaswamy

Production: Rebecca Short

Publisher: Keith Drew

Publishing director: Georgina Dee

Publishing information

Distributed by Penguin Random House
Penguin Books Ltd, 80 Strand, London WC2R 0RL

© 2015 Rough Guides
A Penguin Random House Company
1 3 5 7 9 8 6 4 2

224pp includes index

A catalogue record for this book is available from the British Library

ISBN: 978-0-24125-274-1

Contents

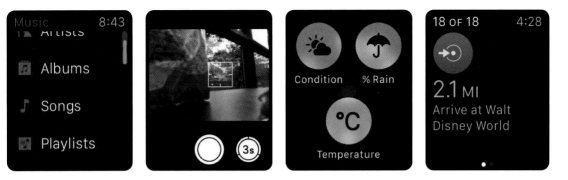

Introduction

Apple has gone and done it again. First Mr Jobs introduced us to the Mac, then the iPod, following up with the iPhone and iPad. And now, Mr Cook brings us Apple's latest game-changing, life-altering, paradigm-shifting device: the Apple Watch.

I've been an Apple guy for going on 20 years now, and I've seen a lot of products come and go. I've been lucky enough to experience several Steve Jobs keynotes in person and have witnessed the launches of some of Apple's most popular products. I have to say that very few have matched the sheer anticipation that preceded the unveiling of the Apple Watch (the iPhone certainly leaps to mind). Apple even had U2 present to help introduce the Apple Watch to the world with a rocking bang. The media hasn't stopped talking about Apple Watch before or since. When Apple Watch finally went on sale (online only at the time), most models were sold out in a matter of minutes – no joke. More Apple Watches were sold in those first minutes of going on sale than all Android Wear watches combined for all of 2014. The question begs asking: is Apple Watch worth all the hubbub?

The simple answer is yes. There's no doubt in my mind that Apple Watch will make a massive impact on how we communicate with one another, how we manage our daily tasks, how we track our fitness goals and levels, how we interact with multimedia and much more than many people have even dreamed of. Like any new technology, Apple Watch has its detractors, just like the iPhone did. You don't hear much from those early iPhone critics these days, and I believe you won't hear too much from the Apple Watch naysayers in the near future, either.

So strap that Apple Watch on, dear reader! You're among the first to adopt a new way of tech life, and I applaud you for it. I've written the *Rough Guide to Apple Watch* for adventurous folks like you, who know the next big thing when they see it. It's my sincere hope that this tome exceeds your expectations as much as the Apple Watch itself will. Welcome to the future!

Acknowledgements

This book would not have been possible, and I would look much less competent, were it not for the entire team behind the whole thing. I know that I will miss some names, and if yours is one of them, please forgive my oversight and know that your work on the *Rough Guide to Apple Watch* is sincerely appreciated.

Very special thanks go to the following teammates: acquisitions editor Janette Lynn (for keeping me on the right track throughout this process), development editor Ann Barton, designer William Thomas and photographer extraordinaire Ralph Anderson! Thank you to Carole Jelen, my agent and friend. Finally, I thank my family and friends for your patience and understanding during the writing of this book. I especially thank my wife, Cindy, and our children: Victoria, Devyn, Emi and Reid. And above all, I thank God who saw fit that we should all walk together on this life's many paths.

CHAPTER 1

Getting to Know

Apple Watch

Anatomy of Apple Watch

The sleek, streamlined design of Apple Watch is designed for easy, intuitive use and has few external parts and pieces.

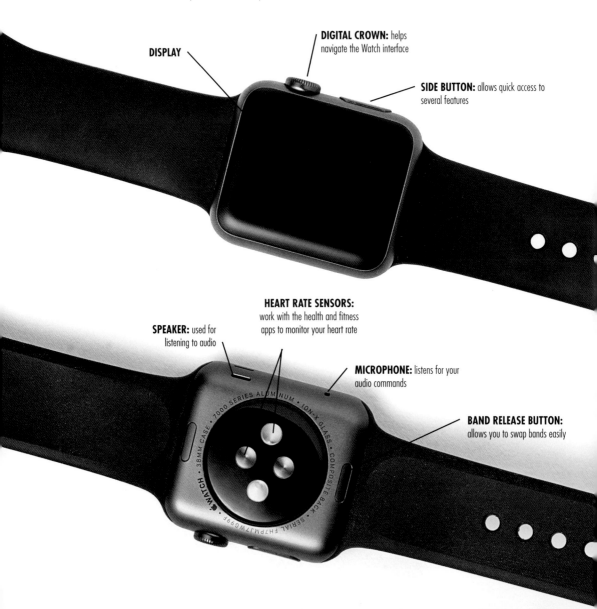

DISPLAY

DIGITAL CROWN: helps navigate the Watch interface

SIDE BUTTON: allows quick access to several features

HEART RATE SENSORS: work with the health and fitness apps to monitor your heart rate

SPEAKER: used for listening to audio

MICROPHONE: listens for your audio commands

BAND RELEASE BUTTON: allows you to swap bands easily

Apple Watch **Models**

Apple Watch comes in three different models, with options that can be mixed and matched to create 38 different variations.

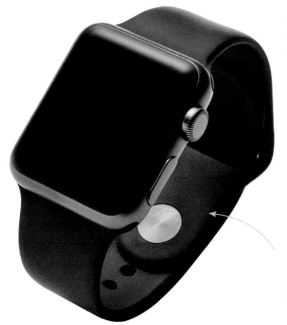

When it comes to the digital functionality of the Apple Watch, every model is created equal. All models include these hardware capabilities:

- Maximum battery life of up to 18 hours
- 8 GB of storage (maximum of 2 GB for music and 75 MB for photos)
- Processor: Apple's S1 chip
- Gyroscope and accelerometer built in
- Taptic engine and heart rate sensor built in
- Water resistant (not waterproof)
- Two size options: 38 mm and 42 mm

The Apple Watch shown in this book is the Watch Sport model with the Space Grey aluminium case and black sport band.

	APPLE WATCH SPORT	APPLE WATCH	APPLE WATCH EDITION
BAND TYPE	FLUOROELASTOMER SPORT BAND IN A VARIETY OF COLOURS	SPORT BAND, LEATHER, STAINLESS STEEL BANDS MAY HAVE POPPERS, BUCKLES, LOOPS OR LINKS	SPORT BAND, LEATHER, STAINLESS STEEL BANDS MAY HAVE POPPERS OR BUCKLES
CASE	ION-X GLASS DISPLAY ANODIZED ALUMINIUM CASE COMPOSITE BACK	SAPPHIRE CRYSTAL DISPLAY STAINLESS STEEL CASE CERAMIC BACK	SAPPHIRE CRYSTAL DISPLAY 18-CARAT GOLD CERAMIC BACK
PRICE RANGE	£299–£339	£479–£949	£8,000–£13,500

What's in **the Box**

Apple is known for its exquisite, minimalist packaging, and the Apple Watch box is no exception. Inside, you'll find everything you need to get started with your new accessory.

Apple Watch and Apple Watch Sport models ship with:
- The Apple Watch
- A magnetic charging cable
- A USB power adapter
- A Quick Start guide
- A band

Apple Watch Edition models ship with the same contents as the Apple Watch and Apple Watch Sport, but also come with:
- A magnetic charging case
- A Lightning to USB cable (for connecting power to the charging case)

Keep Your Packaging

Hold on to the packaging and cases your Apple Watch shipped with in case you need to return it.

The sport band comes in small and large

Apple's USB power adapter has been redesigned to include folding pins.

Charging Your Apple Watch

Your Apple Watch will need to be charged daily in order to keep up with you and your lifestyle. Charging is simple and takes about an hour.

To Charge Your Apple Watch:

1 Connect the magnetic charging cable to a USB port on your computer or use the USB power adapter to connect it to a wall outlet.

2 Lay your Apple Watch's back against the curved side of the round magnetic charger. They will connect magnetically.

If you want to check how far along your charging is, swipe up on the face of the Apple Watch (you may need to enter your passcode), and then swipe right or left to find the charging Glance, which shows the percentage of charge you have available.

If you have an Apple Watch Edition, you can also use the magnetic charging case to charge your Watch.

35% Charged

While charging, your Apple Watch's display will show you how much battery life it has.

The charger connects to the back of the Apple Watch magnetically; there are no charging ports.

A WORD ON BATTERY LIFE
Apple states that the Apple Watch can run on a maximum battery life of up to 18 hours. Your experience may vary, however, depending on the extent to which you put your Apple Watch through its paces. If you're getting dramatically less battery life than you anticipated, by all means contact Apple Technical Support. However, if you're listening to music, texting 100 messages an hour, constantly swiping through your photo collection and checking the weather every five minutes, don't be surprised if you have to charge your Apple Watch more frequently than you'd like.

Interacting with
Your Apple Watch

There are several methods for interacting with your Apple Watch, many of which will feel familiar to those who have used an iPhone or iPad. Even if you're new to Apple products, the intuitive design makes it easy to learn.

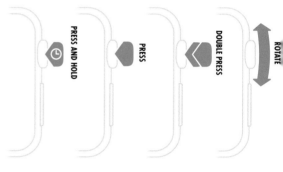

The Display

You can navigate your Apple Watch interface with the simple touch of a finger by swiping, touching and dragging, tapping and pressing the display.

- **Swipe** up, down, left and right to move around on your Apple Watch's display screen.
- **Touch and drag** items to move them around.
- **Tap** icons and menu items to select them or to move down deeper into options and settings.
- **Firmly press** your screen to perform a **Force Touch**. Force Touch allows even more options to be available to the wearer from within certain menus or screens. For example, Force Touch is how you change watch faces, which you'll learn more about in Chapter 3.

The Digital Crown

The Digital Crown is a wholly new input method that is unique to Apple Watch. It can be used in several ways to zip around the interface of Apple Watch.

- **Press** the Digital Crown to go to the Home screen.
- **Double press** (quickly press twice) to open recently used apps.
- **Rotate** to scroll up and down the screen.
- **Rotate** to scroll through lists that are longer than the screen allows.
- **Rotate** to zoom in and out of photos, maps and even apps. For example, with an app centred in the middle of the Home screen, rotate the Digital Crown upwards to zoom right on into the app.
- **Press and hold** to activate Siri, Apple's personal assistant software.

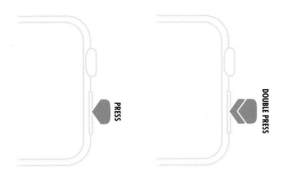

The side button

The side button almost seems like it's trying to hide, since it's nestled unobtrusively under the Digital Crown. It's a great design, but more than that, it's a useful feature.

The side button allows you to do the following:
- Power your Apple Watch on or off, as well as place it into Power Reserve mode.
- Double press to open your Apple Pay cards.
- Lock your Apple Watch. A person must know the passcode to unlock it.

Talking to Siri

In addition to using your finger to navigate your newest timekeeping device, you can also talk to your Apple Watch using Siri, Apple's personal assistant software that recognizes voice commands and questions. Siri will give you real-time responses and can even interact in conversation with you. She's quite a lady, too, as you'll see by her ever-so-polite responses to your queries.

There are two ways to get Siri's attention:
- Wake your Apple Watch and say, 'Hey, Siri.'
- Press and hold the Digital Crown until Siri's screen appears.

What Can Siri Do?

Siri has many capabilities. She can …
- Set alarms, reminders and timers.
- Find your location on a map.
- Check scores of your favourite teams.
- Get local film times.
- Launch apps.
- Search the internet and much more.

To explore the full range of what Siri can assist you with, wake your Apple Watch and say, 'Hey, Siri, what can you help me with?' You'll be rewarded with a very long list of items and actions; just tap one to see examples of questions you can ask her.

English

Español

Français

Français
(Canada)

CHAPTER 2

Setting Up Apple Watch

Why Do You Need an **iPhone**?

The Apple Watch has many capabilities, but it can't do everything on its own. It works in tandem with your iPhone to perform many functions. Having an iPhone is not optional; you must have one to use your Apple Watch.

You Need an iPhone to ...

- Set up your Apple Watch.
- Manage, find, install and uninstall third-party Apple Watch apps.
- Handle tasks your Apple Watch needs to hand off, such as replying to emails or transferring calls.
- Configure many Watch options and features.
- Process data.
- Supply media content.
- Update WatchOS, the Apple Watch operating system.

- Customize notifications on the Apple Watch.
- Act as a gateway to wi-fi networks.
- Back up and restore your Apple Watch's settings, information and apps.
- Organize the location of apps on your Apple Watch's Home screen.
- Add debit and credit cards to Apple Pay for your Apple Watch.

Model Matters

You'll need one of the following iPhone models to use your Apple Watch:

iPhone 6S

iPhone 6S Plus

iPhone 6

iPhone 6 Plus

iPhone 5S

iPhone 5c

iPhone 5

This is a quick way to turn on Bluetooth

Enabling Bluetooth

Apple Watch and iPhone communicate through Bluetooth, a wireless communication protocol that is built in to both devices. Bluetooth is already enabled by default on Apple Watch (as a matter of fact, it can't be turned off), but you may not have it enabled on your iPhone. Bluetooth must be enabled on your iPhone before you can even begin to get out of the gate.

To enable Bluetooth quickly:

1 Swipe up on your iPhone's screen to open the Control Centre.

2 Tap the **Bluetooth** icon (looks like a capital B with wings) in the upper middle of the Control Centre area. If the icon is white, Bluetooth is already enabled.

UPDATING IOS

iOS is the name of the operating system that controls your iPhone. It's necessary to update your iPhone to version 8.2 or later in order to use your Apple Watch. Version 8.2 was the first version of iOS to include the Apple Watch app, which is how you set up, manage and update your Watch.

1 Tap the **Settings** app on your iPhone.

2 Tap **General** and then tap **Software Update**. iPhone checks to see if there is an iOS update available.

3 If your iPhone reports that you have the latest version of iOS, then you're good to go. If not, follow the instructions provided on your iPhone for updating.

Pairing, Setting Up and Unpairing Apple Watch

 To pair your iPhone with Apple Watch means to connect them to one another. This pairing is absolutely crucial to working with Apple Watch. Luckily, it's a cinch to get the devices paired up.

Pairing Your Apple Watch

1 Open the Apple Watch app on your iPhone.

2 Turn on your Apple Watch by pressing and holding the side button until the Apple logo appears.

3 Select a language on the Apple Watch display. Rotate the Digital Crown to scroll through the list.

4 Tap **Start Pairing** on your iPhone and your Apple Watch.

Hold Apple Watch up to the Camera

Align it with the viewfinder below.

5 When the pairing animation appears on your Apple Watch, point your iPhone's camera at it, and line the Watch up in the window provided.

6 If pairing completed successfully, tap **Set Up as New Apple Watch** on your iPhone. If you run into problems, you'll need to pair your Apple Watch manually.

Restoring Your Apple Watch from a Backup

You may have noticed the option to Restore from Backup after your devices were successfully paired. This option allows you to restore settings and information on your Apple Watch if it's been set up and backed up before.

Pair manually if you're having trouble pairing using the camera.

PAIR APPLE WATCH MANUALLY
If you run into a problem pairing the devices, tap **Pair Apple Watch Manually** at the bottom of your iPhone's screen and follow these instructions:

1 Tap the **Info** icon on your Apple Watch's display. You will see the name of your Apple Watch on its display.

2 On your iPhone, tap your Apple Watch's name that appears on the screen.

3 Your Apple Watch will show a six-digit code; enter that code on your iPhone to pair the devices.

Setting Up Your Apple Watch

1 Select the wrist on which you'll wear your Apple Watch and then agree to the Terms and Conditions.

2 Enter your Apple ID username and password. This is a critical step for you to utilize certain features of your Watch.

3 Read and continue through the Location Services and Siri pages, and then decide whether or not to send diagnostics to Apple (your call).

4 Create a passcode for your Apple Watch. It's advisable not to skip this step, as it is vital to protecting your personal information and finances.

5 Determine whether or not to install third-party Apple Watch apps that are available by virtue of you currently having them installed on your iPhone.

6 Your Apple Watch will begin syncing information between itself and your iPhone. This process may take a while depending on how much information is being exchanged. Once syncing is finished, tap **OK** on your iPhone.

Unpairing Your Devices

Why unpair your Apple Watch and iPhone? There are several potential reasons:

- You want to use your Apple Watch with a different iPhone.
- You need to restore your Apple Watch to default settings.
- You want to create a backup of your Apple Watch's settings and information.
- You're troubleshooting connection issues.

1 Open the Apple Watch app on your iPhone.

2 Tap the **My Watch** tab at the bottom left corner of the Apple Watch app.

3 Tap **Apple Watch** in the list of options in the My Watch tab.

4 Tap **Unpair Apple Watch**.

5 Verify that you do indeed want to unpair the devices by tapping **Unpair Apple Watch** at the bottom of the screen.

-91
6:15 PM
Back
Apple Watch
Dwight's Apple Watch
You will need to re-pair with this Apple Watch to use it again.
Unpair Dwight's Apple Watch
Cancel

English
Español
Français
Français (Canada)

APPLE WATCH MUST BE PAIRED TO WORK

You won't be able to use Apple Watch once you unpair it from an iPhone. You must pair your Apple Watch with an iPhone to use it. That doesn't mean the devices must always be in proximity to one another, as you can use Apple Watch for some functions without having the iPhone anywhere around.

Meet the **Apple Watch App**

The Apple Watch app on the iPhone is where you go for everything Apple Watch related. From downloading apps to customizing your settings, the Apple Watch app has you covered.

Your Apple Watch Control Centre

If you've updated your iPhone operating system to iOS 8.2 or later, you have the Apple Watch app. This app has four main tabs, or sections, that allow you to configure your Watch settings, learn more about your Watch and download and manage Watch apps.

Type your search into the search bar and the top App Store matches will populate below.

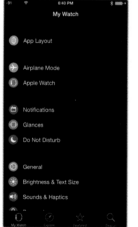

MY WATCH

The My Watch tab is the main hub for configuring your Apple Watch's options. This is where you'll go to adjust settings and preferences.

EXPLORE

This tab includes video tutorials from Apple on how to use your Apple Watch as well as introductions to the new technologies it contains.

FEATURED

This tab highlights some of the most popular and useful apps available from the Apple Watch App Store. See what's new and download what you like.

SEARCH

If you're looking for a particular type of app, go to the Search tab, tap the Search bar at the top of the screen and enter an app name or description.

A QUICK OVERVIEW OF THE MY WATCH OPTIONS

As you scroll through the MyWatch tab in the Apple Watch app, you'll find that there are many options for configuration. The following are some of the options availble on the MyWatch tab.

APP LAYOUT
This option helps you arrange apps on your Apple Watch's Home screen as you like them. It's covered in detail in Chapter 4.

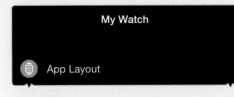

AIRPLANE MODE AND APPLE WATCH
Airplane Mode disables Apple Watch communication features that are incompatible with air travel; see Chapter 4 for more information. Apple Watch shows which iPhone is paired with your Watch.

NOTIFICATIONS, GLANCES AND DO NOT DISTURB
Notifications and Glances are covered in Chapter 8, while more can be found on Do Not Disturb in Chapter 4.

GENERAL AND OTHER SETTINGS
There are many settings for Apple Watch, which are covered in Chapter 4 and elsewhere.

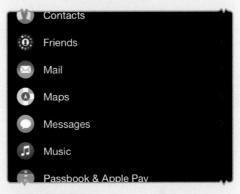

APPLE APP SETTINGS
This section is for configuring options that are particular to the apps that came with your Apple Watch. You'll see more on your iPhone screen than what's shown here.

THIRD-PARTY APP SETTINGS
Third-party apps may need configuring, too, and this is where you access them. The third-party apps shown on your screen will vary depending on which apps you have downloaded.

CHAPTER 3

The
Clock Face
and Time

The **Many Faces** of Apple Watch

Apple Watch incorporates tons of features that the typical watch simply can't perform. One of those features is the ability to modify the clock display with different watch faces and complications.

Watch Face Options

Every Apple Watch comes with 10 different watch face options, from spare and modern to colourful and quirky. Whether you prefer analogue or digital, simple or complex, you're sure to find a face to fit your personality and preferences.

UTILITY

SIMPLE

MOTION

The Motion face is special because the image is always moving. Choose from a blooming flower, a flutterng butterfly, or an undulating jellyfish.

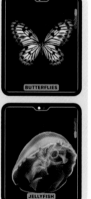

COLOUR

You can change the colour of the Modular face.

MODULAR

MICKEY MOUSE

CHRONOGRAPH

NO EVENTS TODAY

7:58PM

The sunset time is available on some faces.

Complications

In addition to telling the time, watches often have several other hands and miniature faces that add functionality for the wearer. Those other hands and miniature faces are known as **complications**. Apple Watch has many complications from which to choose, including date, moon phase, calendar, alarm, weather and many more. Tapping some complications will open their corresponding apps. For example, tap the Weather complication and the Weather app will open for you.

Battery Life

Weather

Date

Timer

Activity

SOLAR

ASTRONOMY

X-LARGE

Uncomplicated Faces

There are three faces that don't allow for complications: Astronomy, Solar, and X-Large. However, these faces do incorporate other useful features, which are covered later in this chapter.

Selecting and Customizing
Watch Faces

One of the most appealing qualities of the Apple Watch is the ability to choose, modify and save different watch faces to suit your various tastes and needs.

Selecting a Watch Face

Swipe left or right through the options.

1 Press the Digital Crown once to access the Home screen, and tap the **Clock** icon.

2 Force Touch the Apple Watch display (press it firmly with your finger) to open the face selection screen.

3 Swipe left or right until you see the face you want, and then tap to select it.

Adding a Face to Your Selection List

Apple Watch allows you to customize faces and keep your modifications on your selection list for whenever you'd like to use them. For example, you may have two versions of the Simple face, one with nothing but the bare essentials and the other populated with complications you need for various scenarios.

1 From within the Clock app, Force Touch (firmly press) the display to open the selection list.

2 Swipe the faces all the way to the left until you see the **New** button and tap it.

3 Scroll through the list of faces, find the one you want to add and tap it.

Deleting a Face from Your Selection List

Deleting faces keeps your selection screen from becoming cluttered with faces you just never use. If you don't ever use Solar, just delete it. Deleting faces only removes them from the selection list; it doesn't permanently remove them from your Apple Watch. You can always add deleted faces back if you want to.

 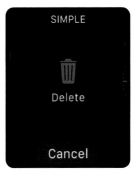

1 From within the Clock app, Force Touch the display to open the selection list.

2 Swipe left or right to find the face you want to delete and swipe up to reveal the Delete trash can icon.

3 Tap **Delete** to remove the face from the list. There is no confirmation needed; once you delete the face, it's gone.

Customizing and Working with Faces

Customizing the face of your Apple Watch is a simple process. However, keep in mind that the customization and complication options will differ depending on the watch face you select.

The Simple face is selected for this example.

1 Press the Digital Crown once to access the Home screen, and tap the **Clock** icon.

2 Force Touch the Apple Watch display (press it firmly with your finger) to open the face selection screen.

3 Swipe left or right until you see the face you want, and then tap the **Customize** button to enter editing mode.

Page dots tell you which screen you're on.

THE CUSTOMIZATION SCREEN
Within editing mode, items that can be customized are highlighted with thin green borders. Note that the customization options will differ depending on the watch face selected. Also, note the small white dots that may appear at the top of the display: the selection of items you can customize changes as you swipe right or left on the display, and these dots let you know which customization screen you're on.

ADJUSTING WATCH FACE COLOURS   EDITING COMPLICATIONS

4 Swipe right or left to select the items and complications you want to customize. Use the Digital Crown to change the settings of each item.

The selected item will have a green border.

The name of the item or complication appears when selected.

DETAIL
Add details such as minutes and seconds by rotating the Digital Crown.

COLOUR
Change the colour of items by rotating the Digital Crown.

COMPLICATIONS
Choose from various complications by tapping to select them.

5 Press the Digital Crown when you're finished making customizations to exit editing mode. Your changes will be saved and your new face shown on the watch face display.

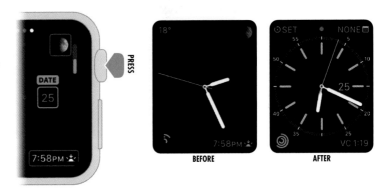

BEFORE　　　　　**AFTER**

Motion, Astronomy and Solar Faces

The Motion, Astronomy and Solar faces don't have many complication options (in fact, only Motion offers complications), but they do have some unique and lovely functions of their own.

Motion

Motion does have two complications you can change: date and animation. Date's a no-brainer, but animation is Motion's claim to coolness. Just be careful to not get distracted by flapping butterfly wings or gliding jellyfish while driving. Rotate the Digital Crown to select one of three animations when Motion is your chosen face.

Tap the object in motion to change it to another.

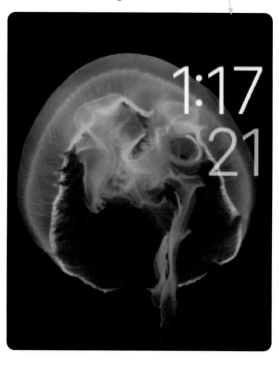

Solar

Solar allows you to see the Sun's position in the sky, and the colours of the face change with the time of day. Rotate the Digital Crown to move forwards and backwards in time, and the Sun will move along the arc of the day.

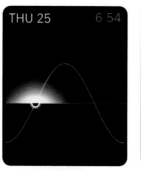

MORNING

MIDDAY

NIGHT

Astronomy

The Astronomy face is beautiful, interactive and educational to boot. You can tell time and get an astronomy lesson at the same time!

Tap the Earth to interact with it.

Tap the Moon to interact with it and see the Moon phases.

This will open the Solar System view.

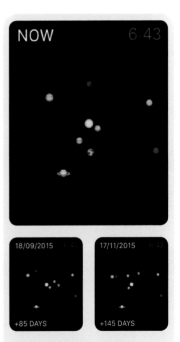

MOON VIEW

Tap the Moon icon in one of the lower corners to make it the active view, and then tap it again to see it as is now. Swipe to spin the Moon or rotate the Digital Crown to see its past, present and future phases.

EARTH VIEW

Tap the globe when Apple Watch is active to see the Earth as it looks now in your area. Swipe the display left or right to spin the globe, or rotate the Digital Crown to see what your part of the world looked like in the past or will look like in the future.

SOLAR SYSTEM VIEW

Tap the Solar System icon to view, and tap the display again to see the planets as they are now in their orbit around the Sun. Rotate the Digital Crown to see where they were orbiting in the past and where they will be in the future.

Setting **Clock Options**

The Apple Watch Clock app has a number of options that you can configure. To adjust these options, you need to go to the Apple Watch app on your iPhone.

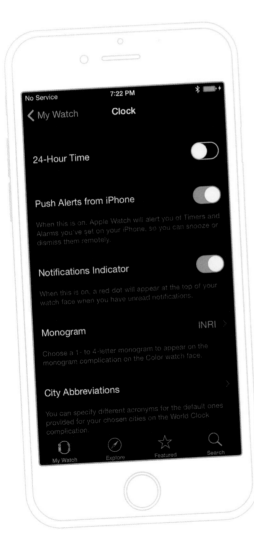

Accessing Clock Options

1 Open the Apple Watch app on your iPhone and go to the **My Watch** tab.

2 Scroll down to **Clock** and tap to open.

24-HOUR TIME
24-Hour Time allows you to display time in 24-hour increments as opposed to the standard 12-hour increments. Toggle the switch to On (green) for 24 hours and to Off (black) for 12 hours.

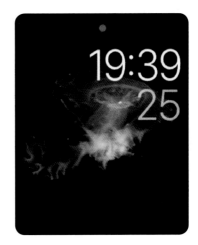

PUSH ALERTS

Push Alerts from iPhone tells Apple Watch to send you alerts for timers and alarms that you've set on your iPhone. You can then use your Apple Watch to keep snoozing or to turn off the alarm or timer. Toggle the switch to On (green) to enable this option or toggle to Off (black) to disable it.

NOTIFICATIONS INDICATOR

Notifications Indicator will display a small red dot at the top of your Apple Watch's display whenever you have an unread notification. Toggle the switch to On (green) to enable this option or toggle to Off (black) to disable it.

MONOGRAM

If you're using the Colour watch face, you may have noticed the Monogram complication right in the centre and wondered how to change it. Tap Monogram and enter a 1- to 4-letter monogram, and it will appear as a complication.

CITY ABBREVIATIONS

City Abbreviations lets you create custom abbreviations for cities in the World Clock app. The World Clock app is discussed later in this chapter.

Keeping Track with the
Stopwatch

The Stopwatch app is a handy feature that can be started and stopped quickly. Stopwatch is extremely accurate and can keep track of multiple laps and times up to 11 hours and 55 minutes.

Using the Stopwatch

Tap to record multiple lap times

Tap to stop timing press again to reset

1 Open the **Stopwatch** app by tapping its icon on your Apple Watch's Home screen.

2 Tap the green button to begin timing.

3 To stop timing, press the red button. To record multiple lap times, press the white button.

Stopwatch Formats

Stopwatch affords several different formats and even plots results as graphs for you so you get a visual representation of timed results. Analogue is the default format for Stopwatch, but it will stay in the last format you used when you close it.

1-DIAL ANALOGUE

3-DIAL WITH LAP SPLITS

To switch from a 1-dial analogue to a 3-dial with lap splits, swipe up on the 1-dial screen. Tap the 3-dial screen to switch back.

ANALOGUE

Force Touch a Stopwatch screen to switch formats. You can select from four.

DIGITAL

GRAPH

HYBRID

Using the **Timer App**

Set your Apple Watch's Timer app whenever you want to be alerted when a certain amount of time has passed. The timer can be set in minute and hour increments up to 24 hours.

Setting a Timer

1 Open the **Timer** app by tapping its icon on the Apple Watch Home screen. (Press the Digital Crown once to get to the Home screen if you're in another app.)

2 Tap the hour button in the middle of the clock and then rotate the Digital Crown to change the number of hours you need. Set to 0 if you only want to measure minutes.

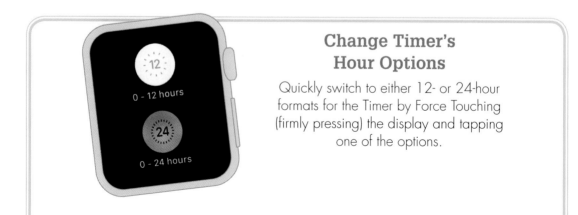

Change Timer's Hour Options

Quickly switch to either 12- or 24-hour formats for the Timer by Force Touching (firmly pressing) the display and tapping one of the options.

3 Tap the minute button to the right of the hour button and rotate the Digital Crown to set the number of minutes you want.

4 Tap the **Start** button to begin the countdown or **Reset** to start over.

5 Tap **Pause** to pause the countdown and tap **Resume** to start it back up. Tap **Cancel** to call the whole thing off.

6 Apple Watch will alert you with sounds and taps on your wrist when the countdown has finished. Tap **Dismiss** to do so.

Setting **Alarms**

Alarms are very important in helping us stay on time throughout our day. Apple Watch has the ability to set multiple alarms to help you keep up with several events.

Setting an Alarm

Switch between AM and PM by tapping the options in the corners.

New

<Edit Alarm   9:13

9:07 PM
Change Time

Never
Repeat

Test
Label

1 Open the Alarms app by tapping its icon on your Apple Watch's Home screen.

2 Force Touch the display. Tap the **New** button.

3 Tap the **Change Time** field on the Edit Alarm screen.

4 Tap the hour button to set hours with the Digital Crown; do the same for the minutes button. Tap **Set** to set the alarm time.

5 When an alarm goes off Apple Watch will alert you with sounds and persistent (yet pleasant) taps on the wrist. Just tap the **Snooze** button on the Apple Watch display (available if you enabled Snooze when setting your alarm) to engage that glorious option, or tap **Dismiss** to cancel the alarm.

Hey, Siri!

You can ask Siri to set an alarm for you by saying, 'Hey, Siri, set an alarm for …' and then the time. The alarm will appear on your list of alarms.

To repeat the alarm on certain days, tap the Repeat field and tap the days you want the alarm to sound.

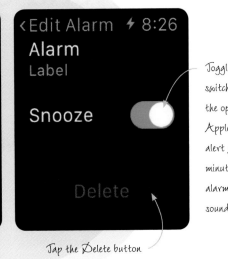

Toggle the Snooze switch if you want the option to have Apple Watch alert you a few minutes after the alarm originally sounded.

Give your alarm a label so that you know what it's for by tapping the Label field. Speak a name for your label and then tap Done.

Tap the Delete button to remove an alarm from the alarms list.

ALARMS ON YOUR IPHONE

If you open the Alarms app on your iPhone, you won't see any of the alarms you've set on your Apple Watch, and vice versa. This is because the alarms on the two devices are completely separate. However, if you have Push Alerts from iPhone enabled in the Clock options within your Apple Watch app's My Watch tab, Apple Watch will alert you of alarms you've set on your iPhone.

MANAGING MULTIPLE ALARMS

If you want to set more than one alarm, just Force Touch the display while viewing the alarms list and tap the **New** button to begin creating it. You can enable or disable alarms by toggling the switch next to them in the alarms list.

Using the **World Clock**

The World Clock app allows you to see the current time in different locations around the world. It's a helpful tool if you're travelling or collaborating with others who reside in different countries.

World Clock on iPhone and Apple Watch

In order to use the World Clock on your Apple Watch, you first need to add cities to the World Clock tab in the Clock app on your iPhone. While the list of cities provided is extensive, it does not include every village and hamlet in the world. You may need to select the nearest large city to your desired location. Once added, the city will be visible in the World Clock app on your iPhone. If you delete a city from the World Clock tab on your iPhone, it will also be removed from the World Clock app on your Apple Watch.

Accessing the World Clock in Apple Watch

The wave shows day and night cycles.

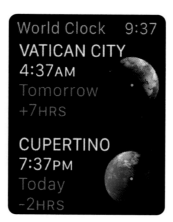

1 Open the **World Clock** app on your Apple Watch by tapping its icon on the Home screen.

2 Scroll through the list of cities by swiping up or down to get a glimpse of the time there, as well as view a map of the city's location.

3 Tap a city to see more detailed information, including sunrise and sunset.

Adding Cities to World Clock

The analogue clock colour is dark when it's nighttime, light for daytime.

Tap the clocks to toggle between digital and analogue.

1 Open the Clock app on your iPhone and tap the **World Clock** tab at the bottom of the screen.

2 To add a city tap the **+** button in the upper right corner of the screen and search for the city you want to add.

Deleting Cities from World Clock

1 To delete a city, tap **Edit** in the upper left corner of the World Clock screen, then tap the **red circle** on the left of the city name.

2 Tap the **Delete** button to confirm.

CHAPTER 4

Configuring
Basic
Settings

The **Settings App** and **My Watch** Options

The basic settings for your Apple Watch can be accessed in one of two ways: through the Settings app on the Apple Watch itself, or through the My Watch tab on the iPhone's Apple Watch app.

Settings App on Apple Watch

 A quick and easy way to make common adjustments to your Apple Watch is through the Settings app. Here you can adjust the time, put your watch in Airplane Mode, set up your Bluetooth devices, change your passcode, manage your accessibility settings and much more. This is your command centre for many of the basic functions of your Apple Watch.

My Watch Tab on iPhone

The My Watch tab on the iPhone's Apple Watch app is another way to adjust your Apple Watch settings. My Watch has many of the same options as the Settings app on the Apple Watch, but some are only found on the iPhone.

Accessing the Settings App

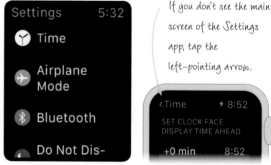

If you don't see the main screen of the Settings app, tap the left-pointing arrow.

1 Press the Digital Crown once to access the Home screen (if you're not already there).

2 The Settings app looks like a gear. Tap it to open the Settings app.

3 Select the item you wish to configure from the list of options.

Accessing the My Watch Tab

1 Find the **Apple Watch** icon on your iPhone and tap to open it. If you don't see one, you need to upgrade the iOS version on your iPhone to at least iOS 8.2.

2 Tap the **My Watch** tab found at the bottom of the screen. Select the item you want to configure.

Adjust the Home screen icon arrangement here.

Options in My Watch are arranged differently than on the Apple Watch.

The My Watch tab is where you access the settings for your Apple Watch.

SOME SETTINGS ARE DEVICE-SPECIFIC

Most of the options covered in this chapter can be configured on both the Apple Watch Settings app and the My Watch tab of the iPhone's Apple Watch app. However, others are specific to one or the other. Look for the icon to indicate whether a particular option is available on the Apple Watch Settings app, on the My Watch tab of the iPhone's Apple Watch App or both.

IN SETTINGS APP   NOT IN SETTINGS APP

IN APPLE WATCH APP   NOT IN APPLE WATCH APP

Time

SETTINGS APP

APPLE WATCH APP

The Time option on the Apple Watch Settings app allows you to set the time displayed on your Apple Watch clock face ahead of standard time.

Why Change the Time?

When you set up your Apple Watch, the time displayed on the clock face will immediately mirror the time on your iPhone. However, if you're perpetually running behind, you may want to set the time displayed on your clock face ahead by a few minutes to help you stay on schedule.

How This Affects Notifications

It's important to note that setting the time ahead will only change the time displayed on your Apple Watch clock face. It does not affect the time set for notifications. Your notifications for reminders, calendar appointments and alarms will still pop up at the actual time. For example, if you've set your clock face 10 minutes ahead, your alarm set for 6:00 AM will sound when your clock face reads 6:10 AM.

You can only set the clock face display time ahead of standard time; not behind.

1 Tap **Time** in the Settings app.

2 Tap **+X min** (X represents the number of minutes being added to the actual time).

3 Turn the Digital Crown to add minutes to the actual time, and tap **Set**.

4 The time an your clock face will reflect the number of minutes you've chosen to add.

Airplane Mode

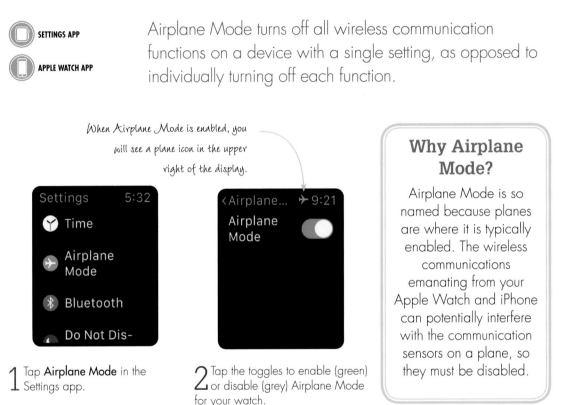

SETTINGS APP

APPLE WATCH APP

Airplane Mode turns off all wireless communication functions on a device with a single setting, as opposed to individually turning off each function.

When Airplane Mode is enabled, you will see a plane icon in the upper right of the display.

Settings   5:32

  Time

  Airplane
  Mode

  Bluetooth

  Do Not Dis-

‹Airplane... ✈ 9:21

Airplane
Mode

1 Tap **Airplane Mode** in the Settings app.

2 Tap the toggles to enable (green) or disable (grey) Airplane Mode for your watch.

Why Airplane Mode?

Airplane Mode is so named because planes are where it is typically enabled. The wireless communications emanating from your Apple Watch and iPhone can potentially interfere with the communication sensors on a plane, so they must be disabled.

AIRPLANE MODE IN MY WATCH
The Airplane Mode option in the My Watch tab of the Apple Watch app does not allow you to turn Airplane Mode on or off on your Apple Watch; it simply allows you to mirror the Airplane Mode setting of your iPhone. With this option enabled, your Apple Watch will go into Airplane Mode only when this function is enabled on your iPhone.

Bluetooth

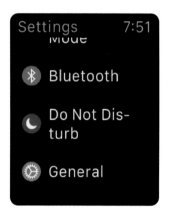

SETTINGS APP

APPLE WATCH APP

Bluetooth is a key technology for your Apple Watch, as it can be used to wirelessly connect many different kinds of devices to your Watch.

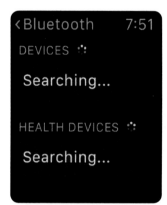

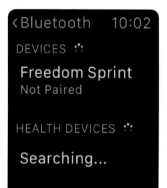

1 Tap **Bluetooth** in the Settings app.

2 Apple Watch immediately begins to search for nearby Bluetooth devices.

3 Once your device is displayed under Devices or Health Devices (depending on the type of device it is), tap the name to connect it to your Apple Watch.

WHAT IS BLUETOOTH?

Bluetooth is a wireless communication technology that is used to allow devices to connect to one another and share information over short distances. As a matter of fact, it's how your Apple Watch talks to your iPhone and vice versa. Many types of devices are being developed to take advantage of the Apple Watch's Bluetooth connectivity. Some commonly used devices include headphones, external heart monitors and hearing aids.

Do Not Disturb

SETTINGS APP

APPLE WATCH APP

The Do Not Disturb function prevents calls and alerts from popping up and sounding off on your Apple Watch, iPhone or both.

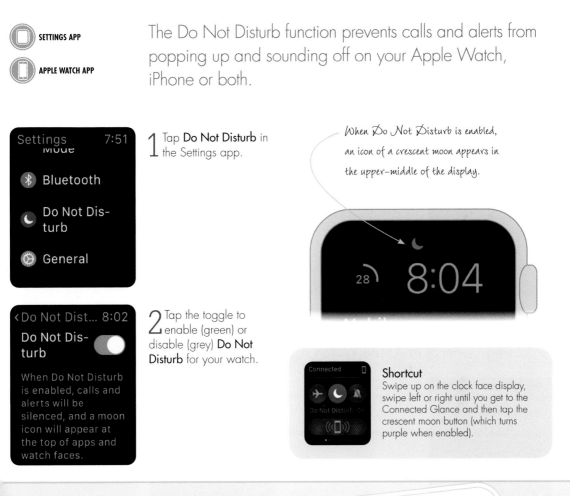

Settings	7:51

Mode

⚫ Bluetooth

🌙 Do Not Dis-
turb

⚙ General

1 Tap **Do Not Disturb** in the Settings app.

When Do Not Disturb is enabled, an icon of a crescent moon appears in the upper-middle of the display.

28) 8:04

< Do Not Dist... 8:02

Do Not Dis-turb ⚫

When Do Not Disturb is enabled, calls and alerts will be silenced, and a moon icon will appear at the top of apps and watch faces.

2 Tap the toggle to enable (green) or disable (grey) **Do Not Disturb** for your watch.

Connected

Do Not Disturb On

Shortcut
Swipe up on the clock face display, swipe left or right until you get to the Connected Glance and then tap the crescent moon button (which turns purple when enabled).

DO NOT DISTURB IN MY WATCH

The Do Not Disturb option in the My Watch tab of the Apple Watch app allows you to mirror the Do Not Disturb setting on the Apple Watch and iPhone. Enabling this setting means when you turn on Do Not Disburb on either device, it will affect the other.

General

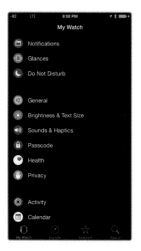

SETTINGS APP

APPLE WATCH APP

Both the Settings app on the Apple Watch and the My Watch tab of the iPhone include a General option, which is where you can adjust many settings.

Accessing General in Settings App

1 Open the **Settings** app and tap the **General** option.

2 Tap the option you wish to configure.

Accessing General in Apple Watch App

1 Open the **My Watch** tab of your iPhone's Apple Watch app, and tap **General** in the options list.

2 Tap the option you wish to configure.

General > **About**

SETTINGS APP

APPLE WATCH APP

About is the place to go for information about your Apple Watch. It can be found in both the Settings app and the My Watch tab for the Apple Watch app.

All About Apple Watch

About in the Settings app is simply a list of information about your Apple Watch and affords no way to make any changes. About in the My Watch tab contains the same information plus a couple of more options.

About contains the following information:
- The name of your Watch
- Number of songs, photos and applications your Watch contains
- Amount of memory capacity and available memory
- The Watch OS version installed
- Your Watch's model and serial number
- Wi-fi and Bluetooth addresses assigned to it
- The SEID number
- Legal information

The SEID (Secure Element ID) number makes Apple Pay an extremely secure payment method.

ABOUT IN MY WATCH

About in the My Watch tab contains a few more options than About in the Settings app.

To change the name of your Apple Watch:

1 Tap **Name** in the options list under About.

2 Enter the name you want to assign your Apple Watch using the onscreen keyboard, and tap **Done** to set it.

To view the Apple Watch User Guide:

1 Scroll to the bottom of the options list in the About tab.

2 Tap **View the Apple Watch User Guide** to open the PDF on your iPhone.

Apple Watch User Guide

General > **Software Update**

SETTINGS APP

APPLE WATCH APP

Software Update is where you download and install the latest and greatest Watch OS updates for your Apple Watch. Install updates whenever they become available.

What Is Watch OS?

Your Apple Watch is a miniature computer, and like any other computer, it needs an operating system to work. Watch OS is simply the operating system Apple has developed for Apple Watch.

When to Update Watch OS

When an update becomes available, you'll see a notification on the My Watch tab. To ensure that your Watch is working at its optimum potential, install updates immediately. The updates are first downloaded to your iPhone and then transferred to the Watch.

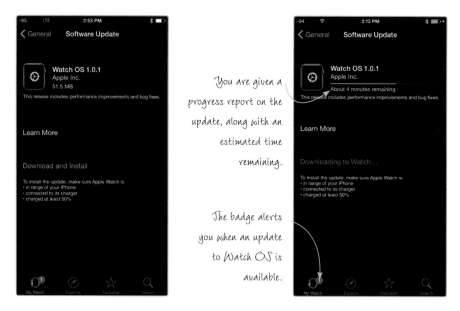

You are given a progress report on the update, along with an estimated time remaining.

The badge alerts you when an update to Watch OS is available.

1 Tap **Software Update** under **General**. Software Update immediately checks for an update and reports what it finds. If there is an update available, tap **Download and Install**.

2 Enter your iPhone's or your Apple Watch's passcodes if prompted, and then tap **Agree** on the Terms and Conditions screen to begin the download.

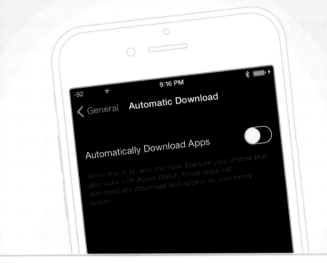

AUTOMATIC DOWNLOAD IN MY WATCH

Automatic Download is another option under the General settings within My Watch. If you have apps installed on your iPhone that also contain apps for your Apple Watch, the Apple Watch versions will be downloaded to the Watch automatically if this option is enabled. Simply toggle the Automatically Download Apps switch to On (green) to enable or to Off (black) to disable.

3 Once the update is finished downloading, you will see a Preparing screen and then an Installing screen on your Apple Watch display.

Your watch will show a progress bar while the update is installed.

4 Your iPhone will report when the update is completed. Simply wait for your Apple Watch to restart itself and you'll be able to continue using it as before.

General > **Orientation** and **Watch Orientation**

SETTINGS APP

APPLE WATCH APP

Your Apple Watch works best if it knows how you're wearing it. You can select your orientation preferences through the Settings app or the Apple Watch app.

Get Oriented

The ever-customizable Apple Watch gives you the choice of wearing it on your left or right wrist and the option of positioning the Digital Crown on the left or right side. For best performance, it's recommended that you tell your Apple Watch how you're wearing it.

To set Watch Orientation:

1 From the **General** menu, tap **Watch Orientation** (on iPhone) or **Orientation** (on Apple Watch).

2 Select your wrist preference (left or right) and Digital Crown preference (left or right).

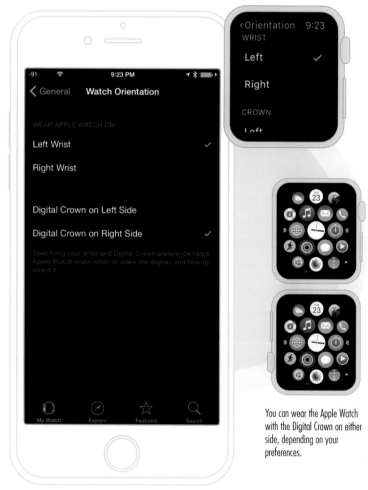

You can wear the Apple Watch with the Digital Crown on either side, depending on your preferences.

General >
Activate on Wrist Raise

SETTINGS APP

APPLE WATCH APP

When enabled, the Wrist Raise function on Apple Watch activates your clock face when you raise your wrist, allowing you to quickly check the time.

Activating Wrist Raise

The Wrist Raise option can only be enabled or disabled from the Settings app on the Apple Watch itself. If the option is disabled, your Apple Watch will only activate when you tap the display or press the Digital Crown or side button, thereby helping Apple Watch to conserve battery power.

To access Activate on Wrist Raise:

1 Open the **Settings** app, tap **General**, and then tap Activate on Wrist Raise.

2 Tap the toggle to enable (green) or disable (grey) Wrist Raise for your watch.

‹Activate on Wris...

Wrist Raise

When this is on, raising your wrist will wake the Apple Watch display. When this is off, you can wake the display by tapping it.

ON WRIST RAISE IN MY WATCH

You cannot enable or disable Activate on Wrist Raise from the iPhone. However, the On Wrist Raise option under the General option in the My Watch tab does allow you to adjust what you see on your clock face when Wrist Raise is enabled.

To access the On Wrist Raise options:

1 Open the Apple Watch app, go to the My Watch tab, tap **General**, scroll down and tap **Activate on Wrist Raise**.

2 Tap **Show Watch Face** or **Resume Previous Activity**.

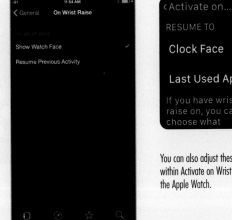

You can also adjust these settings within Activate on Wrist Raise on the Apple Watch.

General > **Accessibility**

SETTINGS APP

APPLE WATCH APP

Apple Watch's Accessibility features make the Watch easier to use so that all wearers can enjoy the full Apple Watch experience.

What is Accessibility?

The Accessibility option under the General tab contains a variety of settings designed to make Apple Watch accessible to anyone. The options here accommodate the needs of those with hearing loss, visual impairments, and motion sensitivity.

You can set many of the available Accessibility options through the General option in both the Settings app on Apple Watch and the My Watch tab in the iPhone's Apple Watch app. However, some Accessibility options can only be configured on your iPhone.

Accessibility Options on Both Devices

VoiceOver

This feature aids the visually impaired by allowing Apple Watch to speak about seleted items on the display. With VoiceOver enabled, you can tap an item to select it and VoiceOver will tell you what the item is and give you verbal instructions on how to use the it. You can double tap a selected item to open or activate it, and swipe with two fingers to scroll.

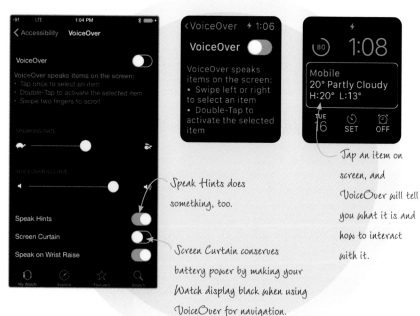

Speak Hints does something, too.

Screen Curtain conserves battery power by making your Watch display black when using VoiceOver for navigation.

Tap an item on screen, and VoiceOver will tell you what it is and how to interact with it.

Zoom

If items on your Watch display seem very tiny, Zoom may be a useful feature. It allows you to double tap the Apple Watch display with two fingers and zoom in or out on the items being displayed. Turn the Digital Crown while zooming to scroll up and down.

Zoom allows you to make smaller interface items larger targets.

Adjust the amount of the zoom by moving the Maximum Zoom Level slider left or right.

Reduce Motion

Some people find that the animation shown when opening and closing apps has a dizzying effect. If you feel this way, you may want to enable the Reduce Motion feature. This option reduces the amount of animation that is shown on the display (for example, when opening or moving icons on the Home screen).

With Reduce Motion enabled, screen transitions will be a subtle fade instead of the zooming effect

On/Off Labels

You've seen how toggling switches on your Apple Watch causes them to display as green when On or grey when Off. However, this colour designation may not be enough for some wearers to fully differentiate between the settings. Enabling On/Off Labels causes On and Off symbols to appear on the toggle switches along with the colours.

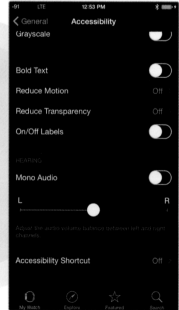

With On/Off Labels enabled, you'll see I when the switch is on, O when it's off.

Accessibility Options in Apple Watch App Only

With Grayscale enabled, everything on the screen appears in black and white.

With Grayscale disabled, the Apple Watch display is colourful.

Grayscale

If a wearer's colour vision is impaired, the Grayscale option will change all colours on Apple Watch to a shade of grey, making it easier to differentiate icons.

Reduce Transparency

This option allows a wearer to see items such as Glances and alerts much more easily by turning down the transparency in their backgrounds.

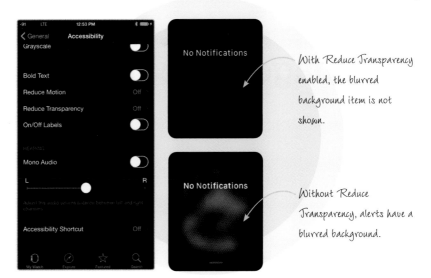

With Reduce Transparency enabled, the blurred background item is not shown.

Without Reduce Transparency, alerts have a blurred background.

TRIPLE CLICK

Mono Audio

Those with hearing impairments can turn off the stereo features of Bluetooth headphones with this feature. Doing so will play all audio channels in both ears, instead of splitting them into stereo channels. This prevents the wearer from missing any of the audio. Dragging the **L** and **R** slider will adjust the volume balance left or right.

Accessibility Shortcut

Apple Watch offers you a quicker way to enable or disable either VoiceOver or Zoom by triple clicking (clicking three times in rapid succession) the Digital Crown. Turn this feature on by tapping **Accessibility Shortcut** in the Accessibility list and then select either **VoiceOver** or **Zoom** by tapping your choice. Simply tap again to disable the choice.

General >
Language & Region

SETTINGS APP

APPLE WATCH APP

The Language & Region options allow you to configure the system language, region format and calendar format for your Apple Watch.

You can quickly find a Region by typing it in the Search bar at the top of the screen.

1 Tap **Language & Region** under **General**. Tap **System Language** to select a different language for your Apple Watch.

2 Tap **Region** to choose a global region format for information such as dates, monetary symbols and the like.

3 If the region you are in doesn't use the Gregorian calendar, tap **Calendar**, and choose either Japanese or Buddhist.

General > **Apple ID**

SETTINGS APP

APPLE WATCH APP

You can't change your Watch's Apple ID, but you can view it by selecting this option.

HOW IS APPLE ID USED?

Your Apple Watch uses your Apple ID for several different tasks, such as keeping track of Apple Pay information, communicating with other Apple Watch wearers, receiving Reminders and other duties. You can only see the Apple ID being used by your Apple Watch in the My Watch tab, not change it. The Apple ID is assigned to the Watch during the initial set-up.

General > **Enable Handoff**

SETTINGS APP

APPLE WATCH APP

Enabling Handoff allows you to switch from your Apple Watch to your iPhone mid-communication.

WHAT IS HANDOFF?

Phone calls, text messages and emails are just some of the communication tasks your Apple Watch is proficient at handling. Sometimes, though, it may be better for a conversation or message to be handled through your iPhone. Handoff is the technology that makes it easy for you to begin communications on your Apple Watch and easily move them to your iPhone. Enable or disable this feature by going to the **General** options and toggling the option On (green) or Off (black) by tapping the switch.

General > **Wrist Detection**

SETTINGS APP

APPLE WATCH APP

This feature activates your Apple Watch screen when you raise your wrist and also locks your watch when you're not wearing it to keep your data secure.

Wrist Detection vs. Wrist Raise

Wrist Detection and Wrist Raise are similar features with one very big difference. The similarity is that both options allow the Apple Watch display to activate when you raise your wrist. The big difference is that Wrist Detection is also a major security function that automatically locks your Apple Watch when you take it off. When you put the Watch back on it detects your wrist and prompts you for your passcode. Enable or disable Wrist Detection by toggling the switch under the General options in My Watch.

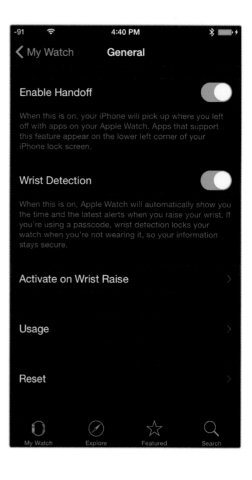

Think Twice Before Disabling Wrist Detection

The benefits of Wrist Detection are plainly obvious. It is highly recommended that you leave this option enabled (it is enabled by default). Without it, it would be very easy for someone to pick up your Apple Watch and use it to gain access to your personal data, including credit card information (if you're using Apple Pay).

General > **Usage**

SETTINGS APP

APPLE WATCH APP

The Usage screen tells you how much space your various apps are using on your Apple Watch, as well as other usage information.

Keeping Tabs on Your Usage

The Usage option is quite straightforward: it basically reports the usage of your Apple Watch to you, which can be helpful if you want to know which apps require the most memory.

Usage shows you:
- The amount of memory you have available and how much you've used.
- How much memory individual apps are using.
- How much time has passed since the last time your Apple Watch was fully charged, how long it was in use and how long it was on standby.
- An estimate of how much Power Reserve your Apple Watch's battery can supply.

A Brief Tip about Power Reserve

Power Reserve mode allows you to still see the time on your Apple Watch, but you're unable to use any other features. Communication is also interrupted with your iPhone, which can help preserve your iPhone's battery if it's low.

-108	6:19 PM	
< General	**Usage**	
Messages		5.0 MB
Calendar & Reminders		3.6 MB
Contacts		2.6 MB
Mail		1.6 MB
Voicemail		No Data
Photos & Camera		—

TIME SINCE LAST FULL CHARGE

Usage	33 Minutes
Standby	3 Hours, 39 Minutes

Power Reserve	> 2 days

Apple Watch will enter Power Reserve when your battery level nears zero.

My Watch   Explore   Featured   Search

General > **Siri**

SETTINGS APP

APPLE WATCH APP

If you'd rather not use the Hey Siri function on your Apple Watch, you have the option of disabling it.

Siri is your Apple Watch's built-in personal assistant, whom we met back in Chapter 1. The option to enable or disable the Hey Siri function for Apple Watch can only be found under **General** in the Settings app on your Watch. Simply toggle the **Hey Siri** switch On (green) or Off (grey). Note that even with the Hey Siri option disabled, you can still access Siri by pressing and holding the Digital Crown.

General > **Regulatory**

SETTINGS APP

APPLE WATCH APP

You'll probably never need to access the Regulatory information for your Apple Watch, but it's there if you want it.

If you take delight in checking regulatory marks on your devices, Apple Watch can scratch where you itch. Simply tap **Regulatory** in the **General** options of the Settings app to see more regulatory info than you can shake a stick at.

General > **Reset**

SETTINGS APP

APPLE WATCH APP

Reset gives you the option to turn back time and return your Apple Watch to factory settings.

Reset lets you, dear Apple Watch wearer, do exactly what it says: reset settings on your Watch. Tap **Reset** in the **General** list of options and you will see three possibilities:

Erase All Content and Settings
This option should give you pause, because it will do what it states. You would select this option if you wish to start all over again with your Apple Watch. You will be prompted once to make sure you are certain you want to go through with the erasure.

Reset Home Screen Layout
Select this option when you're ready to restore your Apple Watch's native apps to their original locations on the Home screen. There's more on this option later in this chapter.

Reset Sync Data
Choose this reset to erase any Calendar items and Contacts that you've synchronized from your iPhone to your Apple Watch. You can resync with your iPhone to restore these items, or update your Apple Watch to better reflect changes in them.

There's No Going Back!

The Erase All Content and Settings reset is permanent. If you choose to go through with this option, **there is absolutely no way to undo it**. You'll be starting from scratch with your Apple Watch afterwards, and you will not be able to recover any of the settings or data you have saved.

Brightness & Text Size

SETTINGS APP

APPLE WATCH APP

The Brightness & Text Size options help to make reading your watch easier by allowing you to dim or brighten your display and adjust the size and boldness of text.

Adjust Brightness

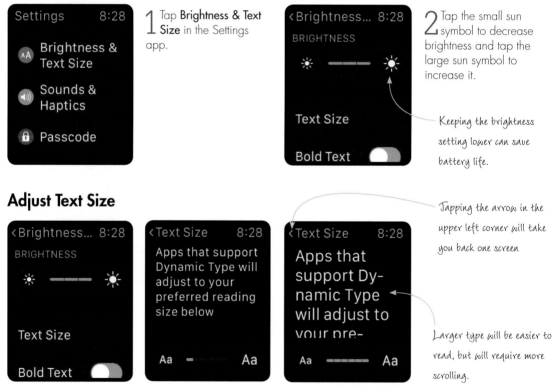

1 Tap **Brightness & Text Size** in the Settings app.

2 Tap the small sun symbol to decrease brightness and tap the large sun symbol to increase it.

Keeping the brightness setting lower can save battery life.

Adjust Text Size

Tapping the arrow in the upper left corner will take you back one screen

Larger type will be easier to read, but will require more scrolling.

1 Tap **Brightness & Text Size** in the Settings app and tap the **Text Size** button.

2 Tap the smaller letters on the left to decrease text size and tap the larger letters on the right to increase it. The text on the screen shows how your settings affect words on the display.

Make the Text Appear Bolder

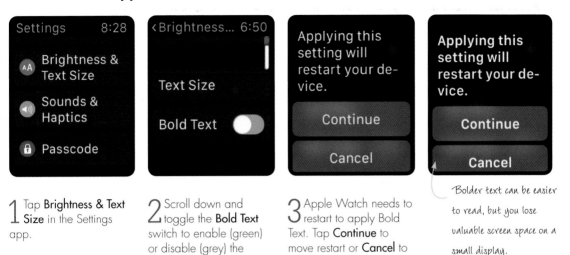

1 Tap **Brightness & Text Size** in the Settings app.

2 Scroll down and toggle the **Bold Text** switch to enable (green) or disable (grey) the option.

3 Apple Watch needs to restart to apply Bold Text. Tap **Continue** to move restart or **Cancel** to exit without changing.

Bolder text can be easier to read, but you lose valuable screen space on a small display.

As in the Settings app, turning on the Bold Text option forces your Apple Watch to restart.

BRIGHTNESS & TEXT SIZE IN MY WATCH

The same options found in the Settings app on your Apple Watch are also available in the My Watch of iPhone's Apple Watch app.

1 Tap **Brightness & Text Size** in the My Watch tab.

2 Adjust the settings for Brightness and Text Size by moving their respective sliders left (decrease) or right (increase).

3 Tap the **Bold Text** toggle switch to turn it On (green) or Off (black).

Sounds & Haptics

SETTINGS APP

APPLE WATCH APP

Apple Watch utilizes sounds and haptics (little taps and vibrations you'll sometimes feel) to alert you when something of interest is happening.

Sounds & Haptics in the Settings App

There may be times when you want to adjust the volume of your Apple Watch alerts or the intensity of the haptics (the taps you'll feel when wearing your Watch). The easiest way to make changes to the sounds and haptics is through the Settings app on your Apple Watch. Simply tap the **Sounds & Haptics** option from within the Settings app to get started.

Tap to access settings for alert volume and haptic intensity.

Adjusting Sounds

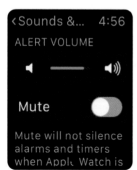

Alert Volume
Tap the speaker icon on the left to decrease volume and tap the speaker icon on the right to increase the volume. As you tap them, you will be given a sample of how much the adjustment affects the alert volumes.

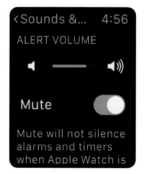

Mute
Toggle the Mute switch to On (green) to silence all alerts on your Apple Watch. Toggle Off (black) to go back to normal sound levels.

Adjusting Haptics

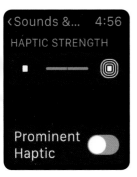

Haptic Strength

Tap the small white dot on the left of the Haptic Strength bar to decrease the intensity of the taps Apple Watch delivers to your wrist, or tap the symbol on the right to increase the strength of the taps. You'll be able to feel the difference the increases and decreases make as you decrease and increase the levels.

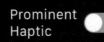

Prominent Haptic

Don't feel the default haptics are enough to get your attention? Enable Prominent Haptic by toggling its switch (green is On) and every haptic you would normally receive is preceded by an additional one. Turn this feature off by toggling the switch to grey.

SOUNDS & HAPTICS IN MY WATCH

1 To access, tap **Sounds & Haptics** in the My Watch tab's list.

2 Decrease the settings for Alert Volume and Haptic Strength by moving their respective sliders left, or increase them by moving the sliders to the right.

3 Toggle the Mute and Prominent Haptic switches to green (On) to enable or to black (Off) to disable.

Cover to Mute

This option is only available in My Watch, not the Settings app. It affords you the ability to quickly mute your Apple Watch by covering the display with the palm of your hand for three seconds. Apple Watch will let you know it has been muted by tapping you on the wrist. Toggle the switch to enable or disable.

Cover to Mute is helpful if you frequently forget to mute your Watch in meetings.

Using a **Passcode**

SETTINGS APP

APPLE WATCH APP

If you wish, you can set a Passcode to prevent others from using your Apple Watch. This is an important security measure that you should seriously consider enabling.

Do I Need a Passcode?

Passcodes are not enabled by default, but you are strongly encouraged to use them. A passcode is required when Wrist Detection is activated, locking your Apple Watch if it is removed from your wrist and only unlocking it when the correct passcode is entered.

Passcode Required

Some features of your Apple Watch, most notably Apple Pay, require that you have a passcode enabled on your Watch for them to work at all.

Set a Passcode

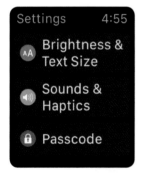

1 Tap **Passcode** in the Settings app.

2 Tap **Turn Passcode On** to assign a passcode.

3 Enter your desired 4-digit passcode on the numeric keypad on the Apple Watch display. (Tap Cancel to back out without setting up a passcode.) You'll be asked to re-enter the passcode to verify it.

Change the Passcode

1 Tap **Passcode** in the Settings app.

2 Tap **Change Passcode**.

3 Enter the current passcode.

4 Enter the new 4-digit passcode.

Turn Passcode Off

1 Tap **Passcode** in the Settings app.

2 Tap **Turn Passcode Off** to disable the passcode feature.

3 Enter the current passcode.

4 If you attempt to turn off passcodes on your Apple Watch when it is storing cards for Apple Pay, you will be given a warning. Should you wish to continue tap **Turn Off**; if not, tap **Cancel**.

Passcode in My Watch

You also have the option of managing your passcode from the My Watch tab in iPhone's Apple Watch app. Although you are using your iPhone, you will still need to have your Apple Watch nearby, as you will use it to enter the passcodes. The Passcode option in the My Watch tab affords the same options as the Passcode option in the Apple Watch's Settings app, along with a few additional capabilities.

KEEP YOUR WATCH HANDY

You can turn Passcode on or off through the Apple Watch app, but it will require you to enter the passcodes on the watch itself rather than in the iPhone.

Selecting these options will prompt you to enter a passcode on Apple Watch.

Turning Passcode off will remove any Apple Pay cards you have stored on your Apple Watch.

Toggle Simple Passcode Off to set a new passcode that is more than 4 digits long.

Use Unique Passcodes

Apple suggests that you use different passcodes for your Apple Watch and iPhone. This is much more secure in the unlikely event that someone else gains access to both of your devices at the same time. It's best to use a unique passcode for your Apple Watch alone.

UNLOCK WITH IPHONE OPTION

This feature allows you to unlock your Apple Watch from your iPhone by using the iPhone's passcode. To enable, it must be toggled On (green) in the My Watch.

If this feature is toggled On (green), that means that the option is enabled on your iPhone in the My Watch tab of the Apple Watch app.

If this feature is greyed out on your Apple Watch, the only way to enable it is by going to the Passcode settings in the My Watch tab on your iPhone.

ERASE DATA OPTION

This security feature causes all data on your Apple Watch to be completely erased should someone enter the wrong passcode after 10 failed tries (after six failed tries, you are locked out for a few minutes). Toggle the switch to On (green) to enable the feature, or toggle to Off (black) to disable.

Use the Erase Data Option Wisely!

Erase Data is a wonderful security feature, but it can cause you unforeseen headaches, too (for instance, if your children start taking guesses at your passcode). If your Apple Watch data is erased, you can pair it with your companion device again and restore from a backup if you have one. The passcode isn't stored in the backup.

Customize Your Watch's
App Layout

SETTINGS APP

APPLE WATCH APP

Apple Watch allows you to customize the layout of the app icons on the display, adding personality and utility to your timekeeping device.

Customizing App Layout on Your Apple Watch

You can continue moving apps around, or press the Digital Crown to exit wiggle mode.

1 Touch and hold an app icon on the display until the other icons begin to wiggle. The app icon you're touching will grow a tad larger than the rest.

2 Drag the icon to wherever you want it to reside, and then remove your finger from the display to drop it in place.

3 When you are done rearranging the icons, press the Digital Crown to exit the customization mode. (Icons will stop wiggling.)

Customizing App Layout from Your iPhone

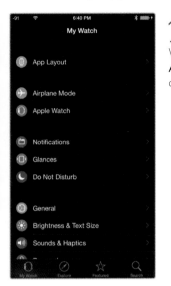

1 Go the My Watch tab of the Apple Watch app, and tap **App Layout** at the top of the options list.

2 Touch and drag an app icon (which will look larger than the others) to its new location. Remove your finger from the iPhone screen to drop the app icon into place.

You can organize the icons in clusters or shapes to help you navigate.

Restoring App Layout

1 Go the My Watch tab of the Apple Watch app, and tap **General** in the options list.

2 Scroll down the next list of options and tap **Reset**.

3 Tap **Reset Home Screen Layout** to restore the app icons to their original positions. Third-party apps will still be visible but may be repositioned.

CHAPTER 5

Phone Calls

Handling **Phone Calls** and **Voicemail**

 Apple Watch uses its built-in speaker and microphone to assist you with answering and placing phone calls, right from your wrist. If you need to, you can hand off the call to your iPhone, too.

Answering or Declining a Call

1 When you receive a call on your iPhone, your Apple Watch will also show the incoming call on its display.

Swipe up to get more options.

Tap the red button to send the call to voicemail.

Tap the green button to answer the call.

2 You can answer the call, or decline it by sending it to voicemail. For more options, swipe up on the display or rotate the Digital Crown.

Tap Send a Message to disconnect the call and send a message to the caller instead.

Tap Answer on iPhone to pick up the call on your iPhone.

Managing a Call

Once you answer a call, your watch will display the call management interface. From here you can adjust the volume, mute the microphone or end the call.

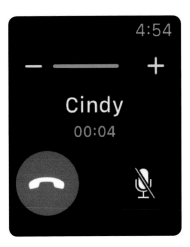

Main Screen
The call management interface displays the name of the caller, and the time elapsed. Tap the red button to end the call.

Volume Control
Tap the + or - buttons to increase or decrease volume. You can also rotate the Digital Crown to control volume; the volume slider will be highlighted by a thin green line.

Mute
Tap the Mute button to mute Apple Watch's microphone so the other party can't hear you. The Mute button turns white when enabled; tap again to unmute the call.

End Call
When the call has ended, your watch will tell you.

SPEAK CLEARLY AND LISTEN CLOSELY
Call quality on the Apple Watch is not the same as on your iPhone. Speak clearly and a bit louder than you normally would, and be prepared to listen closely as the speaker volume isn't such that you'll be able to hear well in some circumstances (if you're walking on a busy city street, for example).

Hand Off to Your iPhone
If you're having difficulty hearing a caller or they're having difficulty hearing you, you can quickly and easily hand the call off to your iPhone.

1 Wake your iPhone (no need to unlock it).

2 Slide up on the Phone icon in the lower left corner of the screen to transfer the call from your Apple Watch to your iPhone.

The phone Handoff icon will appear on your iPhone when you're on a call using your Apple Watch. Slide up to pull the conversation over to your iPhone.

Placing a Call from Friends List

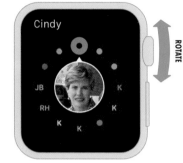

1 Press the side button on your Apple Watch once to open the friends list.

2 To select a friend, rotate the Digital Crown until they are highlighted on the display and then tap their picture or their initial.

3 Tap the **Phone** icon in the lower left corner to place the call.

Placing a Call from the Phone App

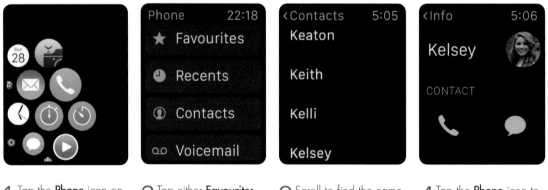

1 Tap the **Phone** icon on the Home screen.

2 Tap either **Favourites**, **Recents** or **Contacts**.

3 Scroll to find the name of the contact you wish to call. Tap the name.

4 Tap the **Phone** icon to place the call.

Working with Voicemail

If you miss a call and the caller leaves a voicemail, an alert will appear on your Apple Watch display. You can address the voicemail directly from the alert or dismiss it to listen to it later.

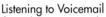

Alert Options
Swipe up on the alert or rotate the Digital Crown to access options to call back, send a message or dismiss the alert.

Voicemail Alert
From here you can tap the play button to hear the message, or swipe up to see other options.

Listening to Voicemail
If you choose to press play on the voicemail alert, you'll see playback controls on the display.

Tap to go back 5 seconds.

LISTENING TO A VOICEMAIL LATER

1 Tap the **Phone** icon on the Home screen.

2 Tap **Voicemail**.

3 Tap the voicemail you'd like to listen to.

Configuring **Phone Options**

Your Apple Watch uses alerts, sounds and haptics to notify you of phone activity. You can choose to use the same alert preferences as your iPhone or customize them for Apple Watch.

Phone in My Watch

To make changes to your phone alert preferences for Apple Watch, you need to start, as you often do, in My Watch. Open the Apple Watch app on your iPhone and go to the My Watch tab. Scroll down to the **Phone** app in the options list and tap to open it.

Tap the Phone icon to get started.

MIRROR IPHONE SETTINGS

The default setting for your Apple Watch phone alerts is **Mirror my iPhone**. With this option selected, your Apple Watch will handle call alerts the same way your iPhone does. For example, if your iPhone makes a sound when you receive a voicemail, your Apple Watch will also make a sound.

Tap the Phone icon to get started.

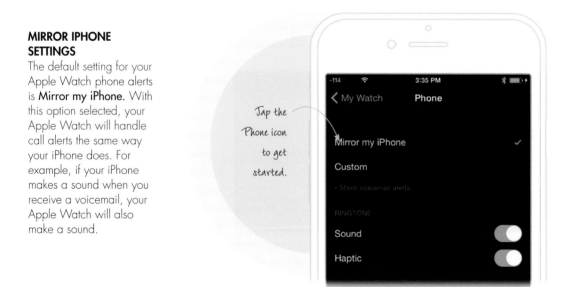

USE CUSTOM SETTINGS

Selecting **Custom** will give you the option of enabling or disabling alerts, alert sounds and alert haptics for Apple Watch independently of your iPhone's alert settings.

Toggle to enable (green) or disable (black) call alerts on Apple Watch.

Ringtone sounds and haptics are set independently of those for alerts.

Toggle the Sound and Haptic switches for alerts to use one or both when alerts do arrive.

The Cover to Mute option is in the Sounds & Haptics menu.

QUICKLY MUTE INCOMING CALLS

Apple Watch also gives you the option of quickly muting incoming calls by placing the palm of your hand over your Apple Watch's display for three seconds. Apple Watch will alert you that sound is muted through a haptic tap.

Enable or disable this feature by opening the Apple Watch app on your iPhone, going to **Sounds & Haptics** in the My Watch tab and toggling the **Cover to Mute** switch.

Managing Your **Friends List**

Add your very favourite contacts to your Apple Watch's friends list so that you can interact with them quickly from within Apple Watch's unique user interface.

Friends in My Watch

In order to make changes to your friends list on Apple Watch, you need to go through the My Watch tab in iPhone's Apple Watch app. From here you can add and remove friends, reorganize your list and disable alerts from specific contacts if necessary.

Tap the Friends icon to get started.

ADDING NEW FRIENDS

1 Scroll down the list of friends and tap **Add Friend**.

2 Browse through your list of contacts to find the person you want to add, or type their name in the Search field to quickly find them.

3 Tap the contact's name to add them to your friends list.

Tap edit to get started.

This icon shows which position the contact is in.

Tap to remove a contact from your watch favourites.

Tap Done when you have finished editing.

Use the grabber to change a contact's position.

Once you tap the red circle, tap Remove to confirm.

EDITING YOUR FRIENDS LIST

Opening the Edit screen in your Friends list allows you to rearrange your contacts and remove contacts from the list.

DISABLING ALERTS FROM SPECIFIC CONTACTS

At the bottom of the Friends menu is the Blocked option. Select this option to block alerts from specific contacts. You're not blocking them from calling you, just blocking Apple Watch from alerting you when they call.

1 Scroll down to the bottom of the Friends screen and tap **Blocked**.

2 Tap **Block Contact**.

3 Browse or search your list of contacts to find the caller you want to block.

CHAPTER 6

Messages, Digital Touch and Mail

Receiving and Reading
Messages

Messaging on the Apple Watch is fun and surprisingly functional, even without a keyboard. When a message alert appears on your Apple Watch, you can respond in a variety of ways.

Message Alerts

Your iPhone will pass messages that it receives to your Apple Watch, provided the two are within Bluetooth range or are on the same wi-fi network, and you'll receive an alert. There are a number of things you can do with the alert.

Tap the alert to open it in Messages.

Tap Dismiss to mark the message as read and dismiss it from notifications.

Press once to leave the message unread but still move it from notifications.

Tap Reply to send a response.

Slide up on the Message icon to transfer the message to your iPhone.

HAND OFF MESSAGES TO YOUR IPHONE
Apple Watch simply isn't equipped to handle drafting long messages or attaching items like photos and videos, so you'll want to hand messages off to your iPhone for that kind of work:

1 Wake your iPhone (no need to unlock it).

2 Slide up on the Message icon shown in the lower left corner of the screen to hand the message from Apple Watch to your iPhone.

Replying to Messages

When you tap the Reply button in a received message, you'll be presented with a variety of ways to respond, from preset responses to animated emojis.

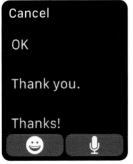

Use a Preset Reply
Apple Watch comes with a variety of present replies that you can use to respond quickly to messages. Tap **Reply** to view and then scroll through the options. Tap a response to send.

If your reply is incorrect, tap Cancel and try again.

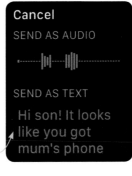

Reply with Dictation
To compose your own message, you'll need to use dictation. Tap **Reply** and then tap the microphone icon. Begin speaking your reply when you see the audio screen. Your reply will appear on the screen once you've finished dictating it.
If your reply is correct, tap **Done**, then choose to send your message as an audio file or as text.

The last screen shows your most recently used emojis.

REPLY WITH EMOJI
Apple Watch comes equipped with both standard and animated emojis to make messaging even more fun. Tap **Reply** and then tap the smiley face button to reply with an emoji. You can swipe right and left to see different emoji styles, and rotate the Digital Crown to see animated variations on the first three emoji styles: Face, Heart and Hand. Force Touch the Face and Heart emoji to change their colours. To send an animated emoji, tap Send. To send a standard emoji, simply tap the emoji itself.

Working with **Photos and Videos** in Messages

Receiving photos and videos adds an extra element of fun to messages. Apple Watch is able to handle attached photos and videos quite nicely.

Viewing Photos in Messages

The pinch-to-zoom feature of iPhone does not work on Apple Watch.

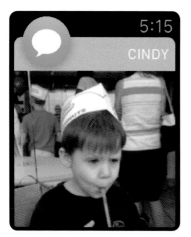

1 Open a message containing a photo on your Apple Watch by tapping the message alert when your Apple Watch receives the message, or by opening it in the Messages app on Apple Watch.

2 Double tap the image with one finger to fit the image on the display, and double tap again to see its original size.

3 Double tap with two fingers to zoom in on the image and do the same trick to zoom back out. Rotate the Digital Crown while zoomed in, or move your finger on the display, to pan around the image.

Playing Videos in Messages

You can quickly jump in 5-second increments.

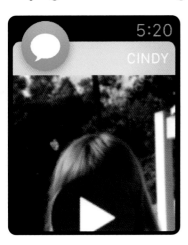

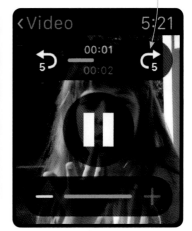

1 Open a message containing a video on your Apple Watch.

2 Tap the video to play it.

3 Tap the Apple Watch display while the video is playing to see the playback controls.

You'll have to save photos and videos using your iPhone.

SAVE PHOTOS AND VIDEOS CONTAINED IN MESSAGES

You'll need your trusty iPhone handy if you want to save photos or videos that you receive in a message. Open the message on your iPhone and save them from there.

Creating **New Messages**

Creating and sending new messages from your Apple Watch is simple. From the Messages app or the friends list, you can use preset replies, send an emoji or dictate a message.

Send a Message from Messages

1 From the Home screen, tap the **Messages** app.

New Message

2 Force Touch the Watch's display and tap the **New** button that appears to begin a new message.

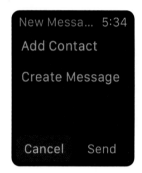

New Messa... 5:34

Add Contact

Create Message

Cancel   Send

3 Tap the **Add Contact** button. Scroll through a list of previous messages to find the contact you want, or tap the blue **Contacts** button to browse your contacts list.

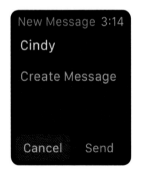

New Message 3:14

Cindy

Create Message

Cancel   Send

4 Tap the **Create Message** button to choose a preset reply or an animated emoji, or to dictate a new message from scratch.

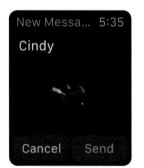

New Messa... 5:35

Cindy

Cancel   Send

5 Tap the **Send** button to send your message or tap **Cancel** if you change your mind.

Animated Emoji on iPhone

Don't worry if the recipient of your animated emoji doesn't have an Apple Watch. As long as they have an iPhone, they'll be able to see the animation in all its glory.

Send a Message from the Friends List

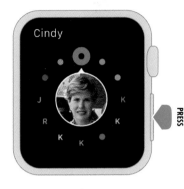

1 Press the **side button** once to open the friends list.

2 Rotate the Digital Crown until the person you wish to contact is displayed. Tap their picture or initial to select them, or just wait a moment and they'll be automatically selected by Apple Watch.

3 Tap the **Message** icon in the lower right of the display to create a new message. If the friend has more than one contact, tap the one you wish to use. Use one of the preset replies, send an emoji or dictate a message.

SEND A DIGITAL TOUCH

Apple Watch's touch display and use of haptics allows it to send a new type of message Apple calls Digital Touch. You can send a series of taps to another Apple Watch wearer, your heartbeat or even create a picture by tracing your finger on the display. Open your friends list and tap the **Digital Touch** button (looks like a hand with a finger pointing up) at the bottom of the display. If you don't see the button, that person doesn't have an Apple Watch.

Tap the coloured dot to change the colour of your drawing.

Button shows that your contact is an Apple Watch user.

DRAW
Draw a picture with your finger and it will automatically send to your friend.

TAP
Tap once or make a series of taps and they will send to your friend.

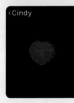

HEARTBEAT
Hold two fingers on the display until your heartbeat is animated. It will be sent to your friend automatically.

Configuring **Messages Options**

Apple Watch uses sounds and haptics to alert you to incoming messages. You can choose to customize your message alert settings for Apple Watch, or mirror your iPhone's message alert settings.

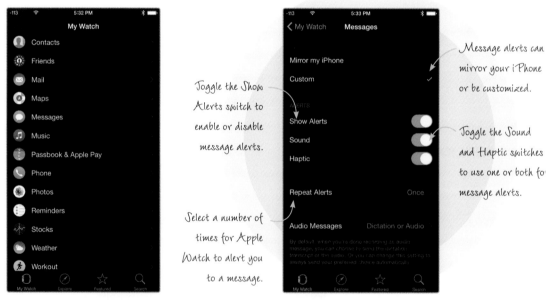

Toggle the Show Alerts switch to enable or disable message alerts.

Select a number of times for Apple Watch to alert you to a message.

Message alerts can mirror your iPhone or be customized.

Toggle the Sound and Haptic switches to use one or both for message alerts.

1 Open the **Apple Watch** app on your iPhone and go to the **My Watch** tab. Find the **Messages** app and tap to open it.

2 If you wish Apple Watch to handle message alerts the way your iPhone does, select **Mirror my iPhone**. To customize how Apple Watch handles message alerts, tap **Custom**.

3 Tap **Audio Messages** and determine if audio messages are to always be dictations, audio files or to allow both options (default) when creating new messages.

When you choose Dictation or Audio, your Apple Watch will ask you each time you use dictation whether to send as audio or text.

The default replies come up when you get a call and choose to send a message.

This setting mirrors the Read Receipts setting for your iPhone by default.

4 Tap **Default Replies** and create custom preset replies for yourself. Just tap a reply in the list and enter your own.

5 Toggle the **Send Read Receipts** switch to enable or disable the feature. Read Receipts tell the people you're messaging that you've read their messages.

Setting Up Apple Watch **Mail**

Mail on Apple Watch works like an extension of the email accounts you have set up on your iPhone. Apple Watch offers a variety of options for customizing Mail preferences.

Mail in My Watch

You can't add email accounts to Mail for Apple Watch; instead, it gets messages from email accounts you've already set up on your iPhone. All the settings and options for Mail on Apple Watch are made using the Apple Watch app on your iPhone. Open the **My Watch** tab and tap **Mail** to get started.

Tap Mail to get started.

Customizing Mail Options

Once you've opened the Mail option in My Watch, you'll find a variety of settings that you can set to your liking. Selecting **Mirror my iPhone** will mimic the settings used for Mail on your iPhone. Selecting **Custom** will provide more options.

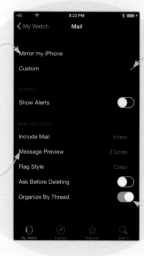

Tap to emulate the settings used for Mail on your iPhone.

Message Preview determines how many lines of a message you can see at a glance.

Tap Custom to have email accounts behave differently on Apple Watch than they do on your iPhone.

This option keeps multiple threads of email conversations together for easier reading.

SHOW ALERTS

Further customize alerts for individual accounts by tapping the account under the Alerts section and then toggle its **Show Alerts** switch On or Off. Toggling the switch to On provides Sound and Haptic switches that you can enable or disable to further customize your interaction.

FLAG STYLE

Customize the look of email flags by tapping Flag Style and choosing either **Colour** or **Shape**.

INCLUDE MAIL

Tap **Include Mail** to specify which inboxes or other email folders should be visible in Mail for your Apple Watch. If you have multiple inboxes and want to see them all, just tap **All Inboxes**.

ASK BEFORE DELETING

If you're anxious about deleting emails accidentally, toggle the **Ask Before Deleting** switch to On (green) to be prompted to confirm email deletions. When you use this option, a single red **Trash** button will appear on the Apple Watch display that you must tap to truly delete the message.

Using **Mail Notifications**

Notifications are Apple Watch's way of letting you know when a new email message arrives. Your notification preferences can be customized in My Watch.

Acting on a Mail Notification

When a new email arrives, a notification will appear briefly on your Apple Watch display. If you ignore the notification, it will fade after a few seconds. If you want to read the incoming email, you can do so by scrolling through the notification. At the end of the email, you'll be provided with other action items, including Mark as Unread, Flag, Trash and Dismiss.

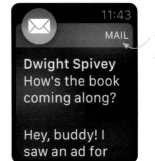

Tap to open in the Mail app.

Read
Scroll through the notification with your finger to read the email, or tap to read in the Mail app. You can read most of the message, but some items, like images, won't display.

Swipe down on your display to see unread notifications.

Ignore
If you've ignored the notification, a red dot will appear at the top of your Apple Watch display to let you know you have unread notifications. Simply drag down from the top of your display to see them, and tap the email you want to read.

Other Actions
Scroll down to the bottom of the email to take further action. (Options may differ depending on the type of email account and the settings you've made for it on your iPhone.)

Customizing Mail App Notifications

1 Open the **Apple Watch** app on your iPhone and go to the **My Watch** tab. Scroll to **Notifications** and tap to open.

2 Scroll to Mail and tap to open. Leave **Mirror my iPhone** selected to use the same notification settings for your Apple Watch that you've already set for your iPhone.

3 Tap **Custom** to customize the notifications for your Apple Watch.

Tap the name of an email account to customize settings for it.

4 Tap the name of an email account to set alerts for the account. Each account can have different alert settings.

Toggle Show Alerts to customize sound and haptic alerts for this account.

NOTIFICATION

NOTIFICATIONS VS. ALERTS

Notifications and alerts sound like very similar things because they are. However, there is a big difference between them. Notifications are reminders of some upcoming event or a notice that you've received a message; they pop up on your Apple Watch so you can respond if you like, and then fade away so you can view them later. Alerts are more urgent notices; they pop up on your Apple Watch and stay there until you acknowledge them with a tap of the display.

ALERT

Using **Mail**

Although you can't use Apple Watch to create new emails or to reply to those you receive, the tricks you can pull off with email on your Apple Watch are pretty great.

Email in the Mail App

1 From the Home screen, tap the **Mail** icon on your Apple Watch.

2 The top left of your display will show the name of the Inbox you've set in your Mail options in the Apple Watch app on your iPhone.

3 Swipe up or down to find the email you want to work with.

Tap a message to read it.

If you have multiple accounts in Mail, the name of the inbox you're viewing will appear here.

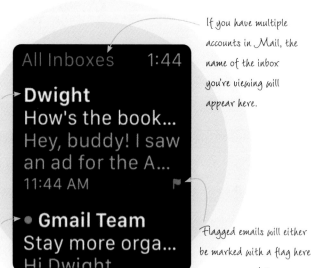

Unread emails are marked with a blue dot.

Flagged emails will either be marked with a flag here or an orange dot.

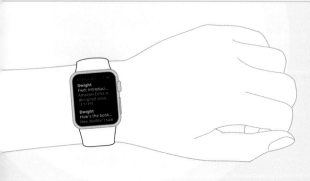

WHY MAIL ON THE WRIST?
Email has been around for quite a long time, and being able to reply to messages, send attachments and the like is old hat to most of us these days. But not having to pull out your iPhone every time an email comes in is in itself justification for getting an Apple Watch if you're someone who is inundated with email during the course of your day. As with everything else Apple Watch, the main reason for Mail on your wrist is sheer convenience.

READING AN EMAIL

Simply tap the email in the list to open. Swipe up or down to read the entire message, if necessary. Force Touch the message (firmly press the Apple Watch's display) to see more options.

Tap the left-pointing arrow to return to the list of emails.

Force Touch the message to see options to flag or unflag, mark as read or unread, or trash.

ACTING ON AN EMAIL WITHOUT OPENING IT

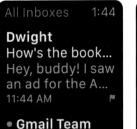

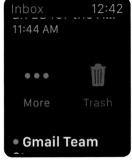

If Ask Before Deleting is enabled you will have to confirm trashing an email.

1 Find the email you want to delete in the list. Do not open the email.

2 Swipe the email to the left. The email preview will shift to the left and reveal action buttons. You can tap **Trash** to delete or **More** for more options.

3 Tap the **More** action button to reveal options to flag or unflag, mark as read or unread or trash again.

Third-Party Email Apps

There are many third-party email apps available in the iOS App Store that you may prefer to use instead of Mail on your iPhone. Many of these apps will work on Apple Watch, too.

Email in Third-Party Apps

If you prefer to use a third-party email app, you may be concerned that you'll have to switch to Mail on your iPhone in order to get email notifications and alerts on your Apple Watch. Allow to me to allay those fears for you, at least in part. I can't speak for every single third-party email app out there, but I know from personal experience that many will work with Apple Watch. Some third-party email apps offer Apple Watch apps, so you can use those as you normally would use Mail on your Apple Watch. And while some do not yet offer an Apple Watch app, you can still work with them using notifications.

Third-party mail apps can offer additional features, like scheduling an email to read later.

Enable third-party app notifications in the Apple Watch app on your iPhone.

ENABLE NOTIFICATIONS FOR THIRD-PARTY APPS

You'll need to be sure that you've enabled notifications for those third-party email apps that don't have Apple Watch counterparts. Open the **Apple Watch** app on your iPhone, go to the **My Watch** tab, tap **Notifications**, find your app under the Mirror iPhone Alerts From section, and toggle its switch to On (green).

Using **Handoff**

One of the coolest features of Apple Watch is the ability to pass tasks on to your iPhone so you can read or work with them more easily. This feature is called Handoff, and it can be used with many native and third-party apps.

Why Handoff?

Some functions, like phone calls and navigation, may be more comfortable or convenient on your iPhone. There are other functions that Apple Watch is simply unable to handle, such as displaying HTML, adding attachments and replying to emails. Handing off these tasks to your iPhone allows you to begin a task on your Apple Watch and seamlessly continue on your iPhone.

Handoff icons will appear in the lower left corner of your iPhone's display.

To hand off email from your Apple Watch to your iPhone:

1 On your Apple Watch, open the email you wish to hand off.

2 Wake your iPhone and you'll notice a Mail icon in the lower left corner.

3 Swipe the Mail icon up to open the email in your iPhone's Mail app.

4 Work with the email as you normally would on your iPhone.

Swipe up on the Mail icon up to access the email on your iPhone.

Some mail messages have to be viewed on the iPhone, not the Apple Watch.

8:18

This message contains elements Apple Watch can't display. You can read a text version below.

CHAPTER 7

Calendar,
Notifications
and Glances

Using **Calendar**

Apple Watch helps you keep track of appointments and invitations with the Calendar app. It syncs with the calendar appointments on your iPhone for seamless scheduling.

Calendar on Apple Watch

Appointments, invitations, practices, concerts, schedules and all other manners of events can make us busy and keep us that way. It's easy to lose track of your day-to-day activities if you don't use a calendar to keep it all together. For years people have relied on their iPhone to help keep them on time, and now Apple Watch can be an extension of that help with its very own Calendar app.

PRESS

1 Press the Digital Crown once to access the Home screen. Find the **Calendar** app (its icon displays today's date) and tap to open it.

2 Calendar will show you today's events upon opening.

OTHER WAYS TO OPEN THE CALENDAR APP
In addition to opening Calendar from the Home screen, you can also access it from a complication on your watch face (if you've set it up) or through Glances.

Swipe up on the watch face to open Glances, swipe right or left to find the Calendar glance and then tap the glance to open the app.

Tap the Calendar complication on your watch face.

Viewing **Events**

Sat
27

The Apple Watch Calendar app sports a crisp and clean interface, and working with your events is simple and intuitive. You can see the day's schedule at a glance, or view complete event information.

In the Today view, tap the left-facing arrow to see the month calendar.

Tap the left-facing arrow in the upper left corner to exit the event details.

1 Open the Calendar app on your Apple Watch. Today's list of events is displayed for you.

2 Swipe up and down to scroll through your events, or just rotate the Digital Crown. Calendar shows you all the events for the next week.

3 Tap an event to see more details. Swipe up and down or rotate the Digital Crown to see more information.

CHANGE THE VIEW FOR CALENDAR EVENTS
The list of calendar events can be viewed in two modes: List and Day. Force Touch (firmly press) the Apple Watch display to see the two modes and tap the one you want to use. If you're viewing in List mode, the option you'll see will be Day, and vice versa.

Adding **Events**

Sat
27

Apple Watch allows you to add events using Siri, but you can only modify events using Calendar on your iPhone.

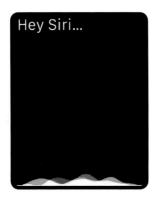

Hey Siri...

1 Say 'Hey Siri' to activate her, or press and hold the Digital Crown until Siri is displayed.

Create calendar event titled first day of school for August 17 8 AM

CALENDAR

First day of school
8:00AM —

2 Tell Siri what you want her to do. For example: 'Create calendar event titled first day of school for August 17, 8 AM.'

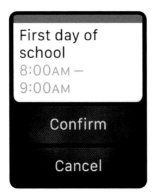

First day of school
8:00AM —
9:00AM

Confirm

Cancel

3 Siri will create your event. Scroll down to **Confirm** or **Cancel** it.

Your event is scheduled for 8 AM August 17, 2015.

CALENDAR

First day of school
8:00AM —

4 If you confirm the event, Siri acknowledges that it has been added to your Calendar.

Configuring **Calendar Settings**

If you want to customize Calendar alerts for your Apple Watch, you'll need to set them up using your iPhone.

1 Open the Apple Watch app on your iPhone and go to the **My Watch** tab. Scroll through the list of options and tap **Calendar**.

2 Toggle the **Show in Glances** switch to On (green) to show Calendar in your Apple Watch's Glances. Toggle to Off (black) to disable the Calendar glance (it's on by default).

3 If you want the Calendar app on your Apple Watch to mimic the Calendar app on your iPhone, tap **Mirror my iPhone** (it's already selected by default). Should you want to customize how your Apple Watch's Calendar app works, tap the **Custom** option.

Toggle to allow Calendar alerts on your Watch.

Which Calendars Does Apple Watch Display?

The Calendar app on your Apple Watch will display any calendars that you are using with the Calendar app on your iPhone. If a calendar isn't displaying on your iPhone, it won't display on your Apple Watch.

Responding to **Invitations**

Sat
27
Whether it's an invitation to a work meeting or a dinner party, the person who sent the invitation would love a response from you so they'll know how to plan.

1 Apple Watch will alert you to an invitation by tapping your wrist and signalling with a tone, and the invitation will show on the display.

2 Swipe the display or rotate the Digital Crown to read the invitation details.

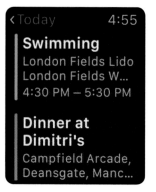

3 Three possible responses are found at the bottom of the invitation.

4 If you tapped Accept or Maybe the invitation is made available for viewing in your Calendar.

Working with an Accepted Invitation

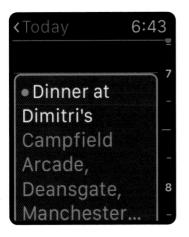

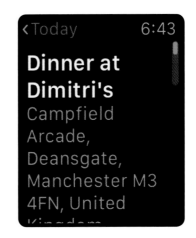

1 Find an invitation in your Calendar and tap to view its details.

2 Swipe up and down or just rotate the Digital Crown to see details of the invitation.

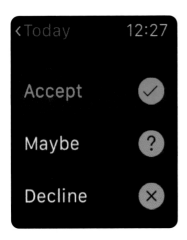

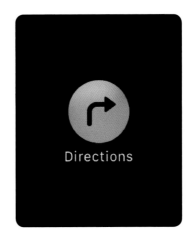

3 Your response to the invitation is highlighted in green, but you can change your response if needed. The person who sent the invitation will be notified of the change.

4 If your invitation came with an address to the event's location, Force Touch the Apple Watch display and you'll see a Directions button. Tap **Directions** to get them from the Maps app.

Notifications

When something important happens, like an incoming message or an upcoming calendar appointment, your Apple Watch will let you know by displaying a notification.

What Are Notifications?

Notifications are messages that alert you when something occurs that you should know about, such as new emails, incoming messages, calendar events and activity alerts. Notifications may be accompanied by a sound or haptic, depending on how you've configured your preferences for specific apps.

Third-Party Apps

Notifications aren't limited to just Apple Watch's native apps. Third-party apps, such as calendar and email apps, can also provide notifications of important events.

Viewing Notifications When They Arrive

Tap the app icon in a notification to open the corresponding app.

1 When a notification arrives, you can work with it right away or dismiss it for later. Raise your wrist to see the notification on your Apple Watch. Swipe or rotate the Digital Crown to read the notification.

2 Some notifications, like Calendar events, allow you to respond by tapping action buttons while others simply let you dismiss them by tapping the **Dismiss** button. Ignore a notification and Apple Watch will allow you to view it later.

Working with Saved Notifications

1 If there are unread notifications, you will see a red dot at the top of your watch face.

2 Swipe down from the top of the Apple Watch display to open the saved notifications list. Scroll through the list by swiping up and down or rotating the Digital Crown.

3 Tap a notification to read and respond to it, if necessary. Swipe to the left on a notification and tap the **Clear** button to clear it from the list.

4 Force Touch the display to see the **Clear All** button; tap to clear all the notifications in the list in one fell swoop.

NOTIFICATIONS IN MY WATCH

You can customize settings for notifications using your iPhone. Open the **Apple Watch** app and go to the **My Watch** tab to start. Find and tap **Notifications** in the My Watch options.

The Notifications Indicator is the red dot that appears on the watch face when unread notifications are available.

Notification Privacy hides details until you tap on the notification.

Select a specific app to configure its notifications.

Glances

Glances are an easy way to access commonly used information quickly without opening individual apps. You can choose the what information you'd like to have available in Glances, including Calendar, Activity, Maps and more.

Opening Glances

1 Swipe up from the watch face screen to open Glances.

The dots at the bottom of the screen are page counters.

2 Swipe right or left to view them.

Tap Power Reserve to turn off all features except time to conserve battery.

APPLE GLANCES

HEARTBEAT
Measures your heart rate

BATTERY
Shows the battery status

NOW PLAYING
Displays the music you're playing

CALENDAR
Displays events for the day

STOCKS
Get a peek at your favourite stock's performance

ACTIVITY
Provides a quick summary of health data

WORLD CLOCK
Shows the time for your favourite city

SETTINGS
Allows for quick configurations of several settings

MAPS
Gives you a map of your current location

Configuring Glances

Find and tap **Glances** in the My Watch options. From here you can add, remove and reorganize your Glances. Scroll down to see the full list of available Glances.

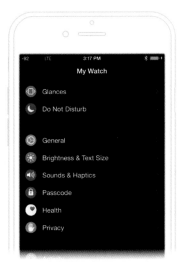

You cannot remove the Settings glance.

Tap the red button to delete a glance, then confirm by tapping Remove.

Tap the green button to add a glance.

Touch and hold a glance's handle to drag it to a new location in the list.

Glances for both native apps and third-party apps are shown.

WHERE'S MY IPHONE?

The Settings glance has a special feature that can help you locate a misplaced iPhone. Swipe up from the watch face screen to open Glances, and then swipe right or left until your find the Settings glance. There you'll see a big blue button. Tap once and your iPhone will sound off with a loud pinging tone. Follow the sound until you locate your missing iPhone.

The iPhone icon indicates that your phone is within range to connect.

Tap to find your iPhone. Your iPhone will sound a tone.

CHAPTER 8

Health and
Activity

The **Activity App**

Apple Watch is a great personal assistant, but it also moonlights as your personal trainer. It comes equipped with the Activity app, which is designed to help you reach your fitness goals.

Activity in Apple Watch

Activity monitors three aspects of your daily activities:

- How often you stand.

- How many calories you burn through movement.

- How long you move at more than a brisk pace.

As you go through your day, the Activity app tracks your progress, which it displays through the use of rings. Each ring corresponds to one of the aforementioned aspects of your activities.

When a ring is closed, you have completed this goal for the day.

Your progress is measured throughout the day.

THE THREE ACTIVITY RINGS

Stand
Your Apple Watch will notify you that it's time to stand up if you've been sitting for too long.

Stand
Stand up for at least 1 minute of every hour.

Move
To reach your movement goals, get up and walk around.

Move
Hit your personal calorie burn goal by moving more.

Exercise
Exercise is measured using your heart rate, movement and speed data.

Exercise
Accumulate 30 min of activity at or above a brisk walk.

Setting Up Activity

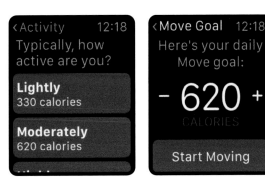

1 Open the **Activity** app by tapping its icon on your Apple Watch's Home screen.

2 Tap each of the four categories to enter your information. Rotate the Digital Crown to sort the information options and tap **OK** to set your selection. Tap **Continue** when you're finished entering the required information.

3 Tell Apple Watch how active you typically are. Select one of the activity description options by tapping. Customize the number of calories you want to set as your daily Move goal by tapping the **+** or **-** buttons.

4 Tap **Start Moving** to begin the Activity app. Your first rings won't look all that impressive, but that will all change as you begin to keep track of your progress.

APPLE TECHNOLOGY FOR HEALTH AND FITNESS

Apple designed Apple Watch to be able to handle the most stressful of workouts, and to track your results as accurately as possible. Several Apple Watch and iPhone technologies help insure this accuracy:

Heart Rate Sensor

The heart rate sensor measures your heart while you're wearing Apple Watch. The information it collects during workouts is used to calculate the amount of calories you've burned and gather other results.

GPS

Your iPhone's GPS provides more accurate times, speeds and distances for Apple Watch so that it can give better reports on your workouts. The GPS information helps more accurately determine how much ground you normally cover, and is especially good to have for calculating workout results when your iPhone's GPS is temporarily unavailable.

Accelerometer

Apple Watch's accelerometer keeps up with information such as how many steps you take and your overall physical movements, and even helps calculate your stride.

Keeping Tabs on
Your Activity

There are a variety of ways that you can keep track of your Activity goals. Quickly view your progress on your Apple Watch, or check your iPhone for more detailed reports.

Activity on Apple Watch

The Activity Complication
Tap the Activity rings complication on your chosen watch face. Not all faces can show the Activity complication.

The Activity Glance
Swipe up from your Clock app's watch face to open Glances. Swipe right or left in Glances to find the Activity glance. Tap the glance to open the Activity app.

The Activity App
Open the Activity app by tapping its icon on Apple Watch's Home screen. The first screen provides you with an overview of all three rings for the day. Swipe down to see a summary report on the day's progress.

ACTIVITY APP REPORTS
The Activity app can provide reports. Swipe left and right to switch between the activities. Swipe down to view your progress in chart form to see when you were moving most, exercising most or standing the most.

Activity on iPhone

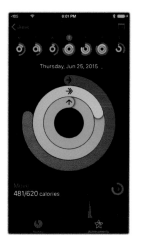

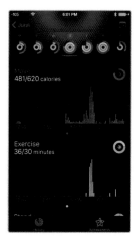

When you open the Activity app, you will see today's progress so far. As on Apple Watch, activity is illustrated with the familiar rings.

Swipe up on the screen to see the graph reports of today's activity for Move, Exercise and Stand.

Swipe all the way down to the number of steps you took throughout the day and how much distance you covered with your feet.

Swipe the bar near the top of the screen that shows the activity rings for the days of the week to move through the days.

Tap the month in the upper left of the screen to see a calendar view of your daily activities, and tap a particular day to see its reports.

Tap the pink calendar icon in the upper right of the screen to be whisked from wherever you may be in the calendar to today's date.

Apple Watch Activity Alerts

Apple Watch alerts you several times during the day regarding your activity levels. You won't have to do anything other than respond to the taps and sounds from your Watch to see alerts, reminders and progress reports for the day's activities.

You'll also recieve weekly summaries that provide an overview of your activity for the preceeding week, as well as a goal assessment that suggests changes for the coming week.

DAILY ALERTS

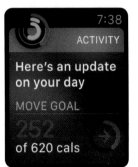

Summary Updates
See the complete update by tapping the alert. Swipe down to the see all of the update, and tap the **Dismiss** button to remove the update from your notifications.

Stand Reminders
These happen whenever you aren't standing as often as you should be. Swipe down to see the entire reminder, and tap the **Dismiss** button to remove the alert from your notifications.

Congratulations
Apple Watch congratulates you when goals are achieved. Swipe down to see the entire message, including a graph report of your achievement. Tap the **Dismiss** button at the bottom to remove the alert from your notifications.

Goal Updates
Individual Stand, Move and Exercise goal updates are provided during the day, as well. Swipe down to read the update. Tap the **Dismiss** button to remove the update from notifications.

WEEKLY SUMMARIES

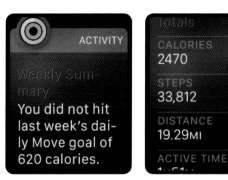

Weekly Summary
These arrive every Monday, so you'll be ready for the new week. Swipe down to read the entire summary. The weekly summary includes a text report of your previous week's activities. You'll also see weekly totals for calories, steps, distance and active time.

Assessing Goals
The Activity app looks at your previous week's activities and recommends a course of action for the upcoming week. If the Move goal recommended by Activity app isn't satisfactory, you can tap the + or - buttons to adjust the Move goal to your liking. Tap the **Set Move Goal** button when finished.

ACTIVITY AWARDS

Apple Watch wants you to know it's pulling for you to reach your health and fitness goals. One way Apple Watch motivates you to meet them is by offering you awards when you've achieved a milestone in your goals.

Award Alerts
When you've earned an award, you'll recieve an alert. Tap the **View Award** button to see the award bestowed upon you by your gracious Apple Watch.

Achievements
To see a complete list of your awards, open the **Activity** app on your iPhone and tap the **Achievements** tab at the bottom of the screen.

Monitoring Yourself with the **Workout App**

The Apple Watch Workout app will help you set goals, track how you're doing, nudge you to success in your workouts and provide reports of how well you did.

Using **Workout**

When you head out to begin a workout, let your Apple Watch know so it can monitor your progress. Selecting a specific workout type will optimize Apple Watch's monitoring capabilities. Whether you're running or biking, using an elliptical or a rowing machine, Apple Watch will take note of your goals and keep track of your workouts.

Improve Accuracy

To get the most accurate heart rate measurement when you use Workout, make sure your Apple Watch fits snugly on top of your wrist. The heart rate sensor should stay close to your skin.

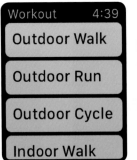

CALORIES
Work out until you have burned a set amount of calories.

TIME
Work out for a specific amount of time.

DISTANCE
Work out until you have covered a specific distance.

OPEN
Work out with no specific goal in mind.

1 Open the **Workout** app by tapping its icon on the Apple Watch Home screen.

2 Select a workout type from the list; rotate the Digital Crown or swipe up and down to move through it.

3 Set goals for your workout by swiping through the available options and set the goal using the + and - buttons. Tap **Start** to begin your workout.

During Your Workout

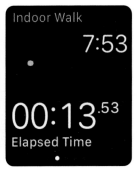 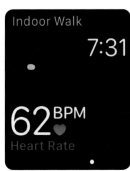 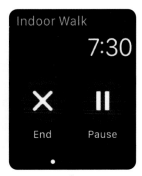

Elapsed Time
View the progress of your workout on your Apple Watch's display.

Progress Indicators
Swipe right and left on the display to see other progress indicators such as Pace, Distance, Calories and Heart Rate.

Pause or End
Swipe to the left to see End and Pause buttons. Tap **End** to stop your workout and **Pause** to pause it; tap **Resume** to continue.

Bring Your iPhone

If you bring your iPhone on an outdoor workout, the Apple Watch will use its GPS to better track your distance, pace and calories burned.

After Your Workout

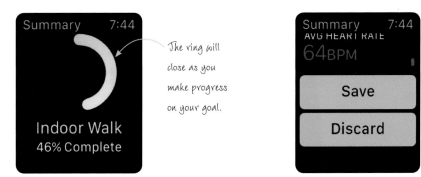

The ring will close as you make progress on your goal.

Summary
The Workout app provides a complete summary of your results. It includes: how much of the workout you completed, the date and time of the workout, the total distance and total time, your active and resting heart rate, the total calories burned and your average pace.

Save or Discard
Tap the **Save** button at the end of the summary to save your workout in your iPhone's Activity app, or tap **Discard** to get rid of the evidence (tap **Discard** again to confirm).

Viewing Workouts on iPhone

1 Open the **Activity** app on your iPhone.

2 Tap the day you completed your workout.

3 Scroll down to the Workouts section (below the Stand report).

4 Tap the workout to see the complete results.

Tap the workout to see details.

Awards earned appear here.

The type of workout, time and goal completion percentage will appear here.

Your total workout time and average heart rate will be recorded.

If you meet your workout goal, the ring will be closed.

Your total calorie burn will be calculated.

Calibrating
Your Apple Watch

Calibration tells your Apple Watch what a typical span of exercise is like for you physically, and it uses that information to provide more accurate workout results.

Get Ready to Calibrate

Calibrating your Apple Watch requires doing a short outdoor run or walk (these activities should be calibrated separately). Prepare yourself as you would for a regular workout, grab your iPhone and Apple Watch and head outside. For optimal calibration, look for a flat, open space and clear skies. Before you begin, make sure your iPhone is held securely, or safely stowed in an armband or waistband.

Warning

Your calibration data is stored on your watch only, so it will be lost if you unpair your Apple Watch from your iPhone.

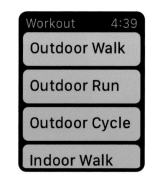

You need at least 20 minutes' worth of data.

1 On your iPhone, go to **Settings** > **Privacy** > **Location Services** to enable **Motion Calibration & Distance** and **Location Services**.

2 Open the Workout app on Apple Watch and select **Outdoor Walk** or **Outdoor Run** for the workout type.

3 Select a goal and tap **Start**, then work out for about 20 minutes. The longer you work out, the more data Apple Watch will gather.

Setting Activity, Workout and Health **Options**

The Activity and Workout apps are configurable so they behave to your liking. The Health options allow you to set your personal statistics in the Apple Watch app. All three apps must be configured on your iPhone.

Setting Activity Options

1 Open the **Apple Watch** app on iPhone and go to the **My Watch** tab.

2 Find **Activity** in the options list and tap to select it.

3 Set the available options as you desire.

Toggle On to make the Activity glance available.

Tap to adjust the frequency of Progress Updates.

Toggle On to receive Award alerts.

Stand Reminders alert you when you've not stood in a while.

Toggle On to receive an alert when you meet a daily activity goal.

Enable to receive a summary of the previous week's activities each Monday.

Setting Workout Options

1 Open the **Apple Watch** app on your iPhone and go to the **My Watch** tab.

2 Find **Workout** in the list of options and tap to select it.

3 Set the two options as you like.

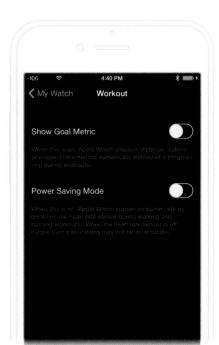

Show Goal Metric
During workouts, Apple Watch will display a progress ring to help you track your goals. The **Show Goal Metric** option causes a numeric list to display instead. Toggle the switch to On if you prefer to see numerals instead of rings during your workout.

Power Saving Mode
This mode turns off the heart rate sensor during your workouts in order to conserve Apple Watch's battery life. Toggle the switch to On to enable this feature.

The heart rate sensors are vital to accurate workout tracking.

A WORD ON POWER SAVING MODE
Your Apple Watch's heart rate sensor is an important piece of the puzzle when it comes to tabulating your workout results. Enabling Power Saving mode in the Workout options turns off the heart rate sensor, which may cause the results of your workout to to be less accurate than they could be. It's best to leave this option disabled unless you're about to begin a workout and your watch is running low on battery power.

Setting Health Options

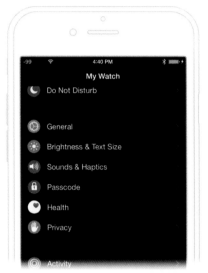

1 Open the **Apple Watch** app on iPhone and go once more to the **My Watch** tab. Find **Health** in the options list and tap to open it.

2 Set or change the following options as needed. Tap the **Edit** button in the upper right corner and tap an option to change it.

3 Tap **Done** in the upper right corner when you have finished making changes.

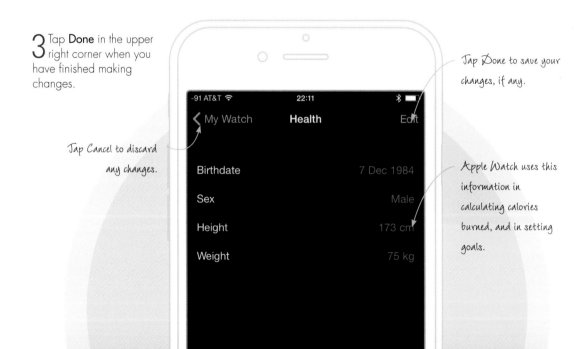

Tap Done to save your changes, if any.

Tap Cancel to discard any changes.

Apple Watch uses this information in calculating calories burned, and in setting goals.

THIRD-PARTY HEALTH AND FITNESS APPS

Apple Watch's Activity and Workout apps are very good at what they do, but there are other third-party health and fitness apps that you may want to use as well. Many of them are tailored to specific kinds of workouts and equipment, which may be a better fit for you. To find these apps:

1 Open the **Apple Watch** app on iPhone and go to the **Featured** tab.

2 Tap **Categories** in the upper left corner and select **Health & Fitness**.

3 Browse the list of health and fitness apps until you find what you're looking for.

Tap Categories to find the Health & Fitness apps.

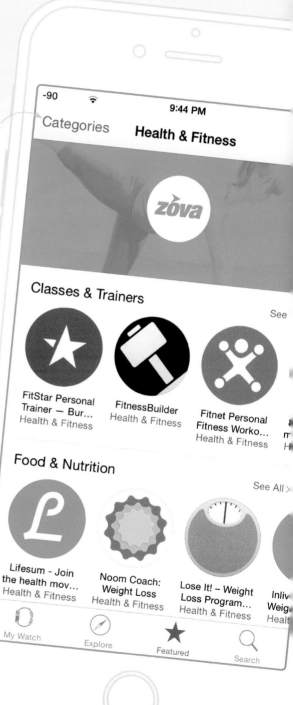

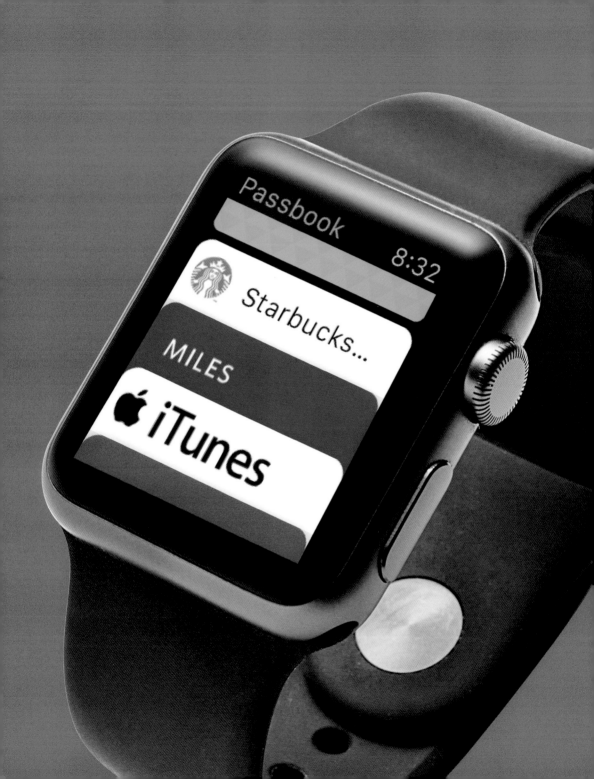

CHAPTER 9

Passbook and Apple Pay

Your Digital **Wallet**

If your wallet is stuffed with credit cards, gift cards and coupons, you may be able to lighten your load by using the Passbook app on your Apple Watch.

Passbook and Apple Pay

Passbook (or Wallet, for iOS9 and later) is an app designed to be a one-stop shop for most items that you might keep in your actual wallet, such as:

- Tickets
- Coupons
- Airline boarding passes
- Store loyalty cards
- Gift cards
- Credit and debit cards

Once passes and cards are stored on your Apple Watch or iPhone, you can use them just by holding your device up to a Passbook or Apple Pay-capable scanner. Passbook works in conjunction with Apple Pay, a secure payment system, to deduct charges from your credit and debit cards. Transactions for passes are simply deducted from the appropriate account. For example, if you use your Starbucks pass to grab a cup of coffee, the amount is immediately subtracted from your Starbucks card balance.

Is Apple Pay Safe?

Apple has gone to great lengths to make Apple Pay as secure as possible, raising the bar high for industry security standards. In fact, handing your card to a cashier or even swiping your card yourself is actually less secure than using Apple Pay. However, no system is foolproof. For more information on how Apple Pay protects you from fraud and theft, visit Apple's Apple Pay security and privacy overview at support.apple.com/en-us/HT203027.

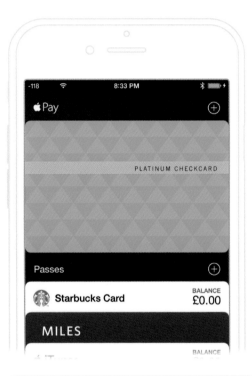

Cards vs. Passes

To make things a bit easier, Apple simply refers to credit and debit cards as 'cards' and all other items as 'passes.'

You can find Passbook-enabled apps in the App Store.

Where Can I Use Passbook?

Many major retailers, airlines, cinema chains and other businesses are utilizing Passbook to provide their customers with a more user-friendly experience. You will see one or both of these symbols at the checkout if a business uses Passbook and Apple Pay.

If you don't see the Apple Pay logo, be sure to check with the retailer.

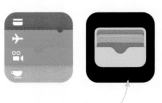

Passbook will be renamed Wallet.

WHAT'S THE DEAL WITH WALLET?

With the release of iOS9, Apple will rename Passbook Wallet. This name change reflects Passbook's ever-increasing role when it comes to purchasing. Over the years, Passbook has evolved from only storing items such as concert tickets and airline boarding passes. These days it can also house your credit and debit cards, and with the change to Wallet, it will also be able to accommodate store credit and loyalty cards. So, since the app is now used more like a wallet than a passbook, the name change just makes sense.

Adding Passes to Passbook

Passes are everything in Passbook that is not a credit or debit card, such as gift cards, coupons or store loyalty cards. These must be added to your iPhone before you can use them on Apple Watch.

Adding Passes to iPhone

In order to add passes to Passbook for Apple Watch, you'll need to add them to Passbook for iPhone first. Once added to Passbook for iPhone, passes are automatically synced with Passbook on your Apple Watch. There are several ways to add passes to Passbook.

 Passbook-Enabled Apps
Many apps will prompt you to add items such as gift cards, boarding passes, cinema tickets and the like.

 Barcode
Scan barcodes that contain pass information using your iPhone's built-in camera.

 Web Browser
Tap a link for passes to add them to Passbook.

 Sharing
Tap the Share icon when viewing passes in Mail, Messages or Safari.

 Email
Click links or attachments that contain pass information.

 Mac
Click links that contain pass information from within Safari or Mail to add them to Passbook. You must be signed in to iCloud.

Syncing Passes to Apple Watch

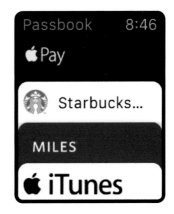

1 Open the Apple Watch app on your iPhone, tap **Passbook & Apple Pay** and tap **Mirror my iPhone** (if it is not already selected).

2 Passes will appear in the stack within the Passbook app on Apple Watch. Scroll through the stack on your Apple Watch by swiping up or down on the screen, or by turning the Digital Crown.

Rearranging Passes and Cards

Passes and cards in Passbook are arranged in a stack according to the order in which they were added, with the most recent at the top of the stack. To rearrange the order of a pass or card:

1 Open **Passbook** on your iPhone. Touch and hold a pass or card to select and move it.

2 Drag the selected pass or card through the list and drop it into the position desired. The new arrangement will also be reflected on your Apple Watch.

Touch and hold the pass or card to move it in the stack.

The order of passes and cards will sync with your Apple Watch.

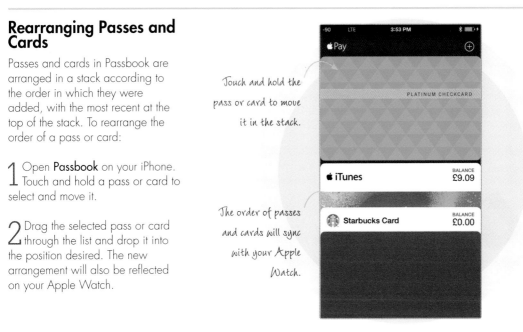

Using Passes
Stored in Passbook

Once you have a pass stored in Passbook, using it is so simple with Apple Watch it almost seems criminal. There are just a few easy steps you need to take to use a pass.

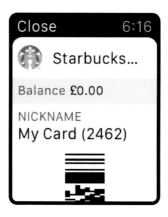

1 Open **Passbook** on the Apple Watch by tapping its icon.

2 Swipe up or down on the screen or turn the Digital Crown to find the pass you want to use, and then tap the pass to open it.

3 Swipe up on the pass to see the bar code; the background colour will fade away in a second.

4 Hold your Apple Watch up to the scanner so it can read your barcode. Your Apple Watch will gently tap you and sound an alert to let you know it's finished communicating the information. Tap **Close** in the upper left of the Apple Watch screen to close the pass if it's still open.

Working with Passes

Passbook on Apple Watch is a quick way for you to access your passes, but you can't do much more than that. In order to work with passes (update them, delete them, etc.), you'll need to use your iPhone.

If you have iOS 9 installed, it will be called Wallet instead of Passbook.

1 Open **Passbook** on your iPhone by tapping its icon.

2 Tap the pass that you want to work with to open it. You may see account information such as your User ID and/or account number, account balances, a barcode and possibly more.

3 Tap the **Information** button on front of the pass (which looks like a small circle containing a lower-case 'i'); locations for this button vary. This action 'flips' the pass so you can view the back.

4 The back of a pass typically provides items such as:

- Information about the pass provider.
- The ability to enable or disable automatic updates.
- Enable or disable **Show On Lock Screen** by tapping the toggle switch. This feature allows your iPhone or Apple Watch to alert you when you are near a location for the pass's provider, such as when you're getting close to a Starbucks.
- Detailed account information.
- Tap **Delete** in the upper left corner of the pass to remove it from Passbook. You'll need to confirm deletion before the pass is removed.
- Tap **Done** to flip back to the front of your pass.

Adding **Credit or Debit Cards** to Passbook

For Passbook to truly be considered your digital wallet, it needs a method of payment. Adding credit or debit cards to Passbook for Apple Watch is an easy process, and also completely secure.

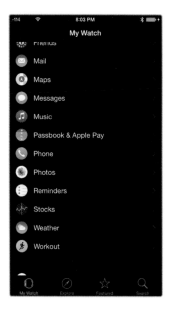

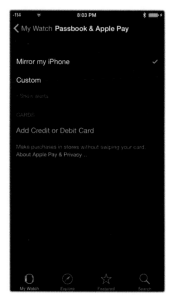

1 Open the **Apple Watch** app on your iPhone. In the My Watch tab, scroll down to **Passbook & Apple Pay** and tap it.

2 Tap **Add Credit or Debit Card** to begin the process.

3 Enter your Apple ID password when prompted and tap **OK**. Tap **Next** in the upper right corner of the Apple Pay screen.

4 If you already have a card on file with your iTunes or App Store account, simply enter the security code found on the back of the card and tap **Next**. You can also tap **Add a Different Credit or Debit Card** should you wish to do so.

5 To add a new card, hold your card in front of your iPhone to use the camera to add it, or tap **Enter Card Details Manually**. On the Card Details screen, either verify the information your iPhone scanned or manually enter the information and tap **Next** in the upper right corner.

6 The steps to verify the card with your bank vary, but you'll need to complete them to continue adding the card to Passbook. In some cases you may be asked to provide information through your bank's iPhone app, you may be asked to call a special number or there may be other modes of verification used.

7 Once verified, the card will show up in the Passbook & Apple Pay section of the My Watch tab and your Apple Watch will alert you that the card is now ready to be used with Apple Pay.

Default Cards

The first card you add is your default card for Apple Pay purchases. You can change your default card quite easily by tapping Default Card in the Passbook & Apple Pay section of the My Watch tab and tapping the card you want to use as default.

Using **Apple Pay**

Apple Pay is a quick and easy way to make in-store purchases without ever opening your wallet. Payment readers at the checkout communicate wirelessly with the Apple Watch, making transactions fast and simple.

Look for the Logo

There are many different manufacturers of Near Field Communication (NFC) readers, and readers may look different from store to store. Just remember to be on the lookout for the Apple Pay logos at or near the checkout, and sometimes near the entrance to the store.

No iPhone Needed

There are several tasks that Apple Watch can perform without need of an iPhone, and using Passbook is one of them. Just make sure that whatever pass or card you want to use is loaded on the Apple Watch beforehand.

Making a Purchase Using Apple Pay

DOUBLE PRESS

1 At checkout, double tap the button on the side of your Apple Watch to access your default card. You can also open Passbook to access the card. If you do go through the Apple Watch Passbook app, select the card you want to use and then double tap the button to make the card active for the purchase.

2 Your default card appears on the Apple Watch screen. If you want to use a different card, just swipe to the left or right until you find it.

WHAT IS NFC?

Near Field Communication (NFC) is the technology employed by Apple for the Apple Pay system. Your iPhone and Apple Watch are both outfitted with NFC technology, as is the reader you find in stores that support Apple Pay. NFC establishes a radio connection between two devices at very short distances (10 cm, or 3.9 inches, or less), allowing them to communicate data to and from one another. NFC has been used in smartphones for years and is a safe and reliable connection type for temporary transmissions of information. NFC is used for communications other than financial transactions as well, such as gaming and even healthcare functions. You can find much more about NFC at nfc-forum.org.

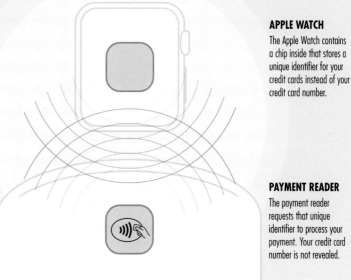

APPLE WATCH

The Apple Watch contains a chip inside that stores a unique identifier for your credit cards instead of your credit card number.

PAYMENT READER

The payment reader requests that unique identifier to process your payment. Your credit card number is not revealed.

Payment readers will often have a symbol like this, showing they're NFC-compatible.

Even if it takes several attempts to get a payment verification, your card will not be charged multiple times. You will only see a verification when a transaction is successful.

3 Hold the Apple Watch near the reader to begin payment. Your Apple Watch will connect wirelessly to the reader by simply placing it directly in front of the reader.

4 Once payment is made, your Apple Watch will alert you with a sound and a gentle tap. You will also see verification on the screen. If your watch doesn't alert you or provide a verification, repeat the procedure until you're successful.

CHAPTER 10

Camera
and Photos

Syncing Photos

We all love to show off our best photos of the people and places that make our lives a joy. Apple Watch makes it easy to view your favourite shots with the Photos app.

The Photos App

Your Apple Watch can't take photos on its own, but you can sync photos from your iPhone to view on your Apple Watch. To sync photos, you'll need to specify an album. By default, Apple Watch is set up to display photos you've marked as favourites in your iPhone's Photos app, but you can change this default setting to any album you choose.

MARKING PHOTOS AS FAVOURITES ON YOUR IPHONE

 FAVOURITE

 NOT FAVOURITE

Browse your photos to find a picture you want to mark as a favourite.

Tap the **Favourites** icon at the bottom of the screen. The Favourites icon is in the shape of a white heart if the photo is not marked as a favourite, but is a blue heart when it is marked as a favourite.

Photos you mark as favourites will automatically appear in the Favourites Album.

Your favourite photos will also be synced across to your Apple devices if you use iCloud.

Selecting an Album to Sync with Apple Watch

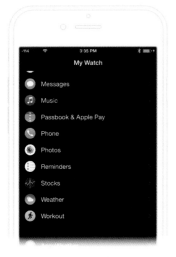

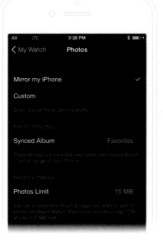

1 Open the **Apple Watch** app on your iPhone. Find **Photos** in the list of options in the **My Watch** tab a tap it.

2 The Synced Album option is set to Favourites by default. Tap **Synced Album** to see a list of available albums you can choose to sync with your Apple Watch.

3 Tap the album you want to use as your Apple Watch's new default. (You can only have one album synced at a time.)

ADDING PHOTOS TO A NEW ALBUM ON IPHONE

1 Open the Photos app on your iPhone. Tap Albums at the bottom of the screen (if you're not already there). Tap the + button in the upper left corner of the screen to create a new album, give the album a name and tap **Save**.

2 Browse the photos on your iPhone and tap the ones you want to add to your new album (a small blue circle with a tick will appear on a selected photo). Tap **Done** in the upper right corner when you've finished selecting photos to populate your new album.

Viewing Photos

Viewing photos on your Apple Watch is very intuitive, and you'll soon be a pro at showing the world your favourite memories with a flick of your wrist.

Viewing Photos on Your Apple Watch

Now that you have photos on your Apple Watch, it's time to show them off to anyone you can! The Photos app makes it easy to organize and view your photos. The zoom feature allows you to see your photos in detail, even on a small screen.

You Don't Need iPhone to View Synced Photos

Once you've synced photos with Apple Watch, you don't need the iPhone nearby to view them; synced photos are stored on your Apple Watch.

Zoom in and out of the collage with the Digital Crown.

Tap and drag on the display to move around in the collage.

1 Tap the **Photos** icon on your Apple Watch's Home screen to open it. The Photos app shows you some of your photos in a collage.

2 Tap a photo to view it. The selected photo will fill the screen.

Zoom out of a photo, and back into the collage, using the Digital Crown.

You can't pinch to zoom, like on iPhone. Only the Digital Crown works for zooming.

Double tap a photo to zoom in or out.

When zoomed in, tap and drag the display to look around the photo.

Toggle Off if you don't want to receive notifcations about shared photos.

MANAGING ICLOUD PHOTO SHARING ALERTS

If others are sharing photos with you using iCloud's Photo Sharing, you get notifications (alerts) whenever they add new photos to their Photo Stream. You can configure your iPhone to either receive these alerts or not. You also have the option to receive those alerts on your Apple Watch.

1 Open the **Apple Watch** app on your iPhone. Find **Photos** and tap it.

2 Choose whether to allow your Apple Watch to mirror your iPhone's settings for these alerts, or tap **Custom** to decide whether or not your Watch gets the alerts independent of your iPhone.

Managing **Photo Storage**

In order to balance the storage demands of music, photos, apps and other items, Apple Watch allows you to set limits on the space used by your photos.

Allocating Storage to Photos in Apple Watch

Your Apple Watch only has so much space to use for items like photos and music, so you should be judicious with it. You can limit the amount of storage space your Apple Watch is allowed to use for storing photos from within the My Watch tab in iPhone's Apple Watch app.

By default, 100 of your favourite photos are synced to your Apple Watch.

1 Open the **Apple Watch** on your iPhone and go to the **My Watch** tab. Find **Photos** and tap it.

2 Find the **Photos Limit** option at the bottom of the screen and tap to access it.

3 Tap a storage limit to set it as the default for your Apple Watch. Available limits are:
- 25 photos = 5 MB (megabytes)
- 100 photos = 15 MB (default)
- 250 photos = 40 MB
- 500 photos = 75 MB

RUNNING OUT OF SPACE?

One way to gain storage space on your Apple Watch is by limiting the number of photos it is able to store. It's important to note that the amount of photos listed is approximate. The concrete limit is actually the amount of megabytes listed. For example, let's say you selected the 100 Photos/ 15 MB option. It's conceivable, based on the size of your photos, that you could reach the 15 MB limit but have fewer than 100 photos stored, or perhaps more than 100 photos if their sizes are small. Another way to free storage space is to limit the amount of songs you can copy to your Apple Watch.

HOW MANY PHOTOS ARE ON MY APPLE WATCH?

You can check the number of photos stored on your Apple Watch without counting them manually.

1 Open the Settings app on your Apple Watch. Tap **General** and then **About**.

2 Scroll down until you see Photos: the number of them stored on your Apple Watch will be listed.

Using the **Camera Remote** App

Apple Watch comes with a nifty little app called Camera Remote, which does the job its name implies: it allows you to remotely take pictures using your iPhone's camera.

When to Use Camera Remote

The Camera Remote app can be helpful in any situation where it's difficult to physically hold your iPhone to take a picture, for example, it may come in handy for …

- Group photos. Now you can take part in pictures, instead of just being the shutterbug who's always left out.
- Selfies, especially if you want something in the background that you can't capture just by extending your arm.

- Maintaining focus. Mounting your iPhone with a tripod can make it cumbersome to take a picture because pressing the buttons can cause the iPhone to get out of the position. Camera Remote eliminates the need to touch your iPhone once it's positioned.
- Low lighting conditions, when just the slightest of movements, such as the pressing of the buttons on your iPhone, can cause blurry photos. Camera Remote allows your iPhone to remain still, thereby producing better low light images.

APPLE WATCH MUST BE WITHIN RANGE OF YOUR IPHONE

In order to use Apple Watch as a camera remote, it must be within Bluetooth range (about 10 metres, or 33 feet) of the iPhone it's partnering with. If the Bluetooth connection is spotty your remote picture taking experience will be, as well.

You could be about three elephants away from your iPhone and still take a picture.

1 Position your iPhone as needed to take your picture. You do not need to open the Camera app on your iPhone, although you can open it to get it in position if you'd like.

Make sure the iPhone is properly supported in position, so it doesn't fall.

The iPhone doesn't need to be unlocked for Camera Remote to work.

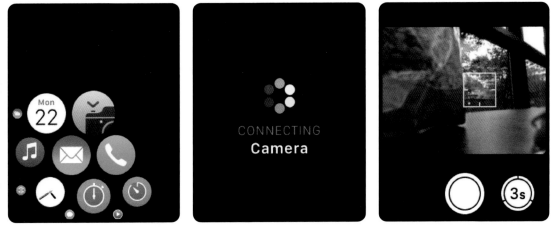

2 Tap the **Camera Remote** app on your Apple Watch. It will begin searching for a connected camera.

3 Camera Remote will automatically open the Camera app on your iPhone once connected. Apple Watch will alert you if the Camera app on the iPhone unexpectedly closes.

4 Your Apple Watch's display will show you what the viewfinder on your iPhone sees, which helps to better place your iPhone for a great shot. This comes in handy when the iPhone is in a position that makes it difficult for you to get a look at what it's seeing.

TAKING A PHOTO

The Shutter button allows for one-time snapshots. Tap the **Shutter** button (the round white button in the bottom centre of your display) to take a picture.

Tap an area in the viewfinder to adjust the focal point.

You can also press the side button to take a photo.

Press the Shutter button to take a photo.

Press the Timer button to shoot a series of three photos after 3 seconds.

USING THE TIMER

1 Tap the **Timer** button (displays the number 3) to set a 3-second time in motion.

2 Three seconds gives you plenty of time to get your subjects into position. Through a series of beeps, flashes and onscreen countdowns, both your Apple Watch and iPhone will give indications that the shots are about to happen.

3 Your iPhone will take three shots in rapid-fire succession, helping to ensure that at least one of them is what you want to capture.

Both your iPhone and A Watch will show the count.

You can tap the small inset to review photos taken during the current remote session.

Taking **Screenshots**

There may be times when you want to capture what's displayed on your Apple Watch screen and save it for later reference. Apple Watch has you covered with the ability to take screenshots.

Screenshots on Apple Watch

A user is able to take screenshots on most digital devices, especially computers and smartphones, quite easily with a special combination of button pushes (some of which require very nimble fingers). Apple Watch also allows you to take shots of its display, which can come in handy in a variety of situations.

Maybe a technical support team you're working with needs to see a problem as it happens, or perhaps you just want to capture an especially nice Digital Touch from a loved one. Whatever the reason, you can take a screenshot quickly and easily with Apple Watch.

The display will flash white when you take a screenshot.

PRESS SIMULTANEOUSLY

Screenshots will appear in your iPhone Camera Roll and in your iCloud Photo Library, if you use it.

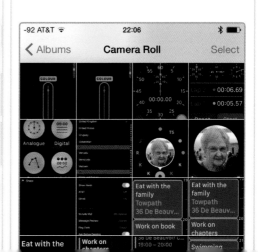

To take a screenshot of your Apple Watch, press the side button and the Digital Crown simultaneously. The display on your Watch will flash white and a sound like that of a camera shutter will be heard (if you haven't muted your Watch).

Music

The **Music** App

As its name implies, the Music app allows you to listen to music on your Apple Watch. You can browse your iPhone music library if it's within range, or store a playlist on the Watch itself.

Playback and More

When your iPhone is nearby, you'll be able to control the playback for your iPhone music files using the Music app on Apple Watch. You can also store a playlist with up to 2GB of audio files on your Apple Watch, making it easy to listen to music while working out or any other time you don't want to be encumbered by your iPhone. Just make sure to invest in a pair of Bluetooth headphones; you won't be able to play your stored playlist unless your Watch is connected to headphones via Bluetooth.

Accessing Music

From the Home Screen
Press the Digital Crown and tap the Music icon on the Home screen.

From Glances
Swipe up on the screen from within the Clock app to see your Glances. If you don't immediately see the Music glance, which shows the playback controls and the song currently playing, swipe left or right through your Glances until you see it.

Getting Around in Music

The Music app is not only functional, it's also fun to use. The menus are simple yet aesthetically pleasing, making it a wonderful option for browsing your music library without the need to fumble around for your iPhone.

Navigating your audio collection:

- Tap any of the main menu items to quickly and easily browse your audio collection.
- Drill down into your collection by tapping individual items.
- To back out of selected items, tap the upper-left corner of the screen. A white left-pointing arrow appears next to the name of the selected item to direct you.
- Tap the name of song or audio file to begin playing it.

Shows the name of the artist of the currently playing song or audio file.

Browse audio by the artist's name.

Browse music and audio by song titles.

Browse audio files according to the album's name.

A green scrollbar appears when using the Digital Crown to browse files.

Peruse the collection of audio playlists that you've created.

Tap the white left-pointing arrow to back out of selections.

Tap an album to view its contents.

Scroll down to see more albums.

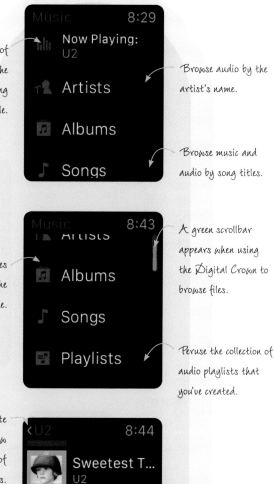

Faster Scrolling

It's easier to use the Digital Crown if you're browsing a long list of audio files, as opposed to frequently swiping up, up and still up again (or down, as the case may be).

Controlling **Music Playback**

Listening to music on your Apple Watch is simple and intuitive. Basic controls are shown on the Now Playing screen, while other options can be accessed by Force Touching the display.

Basic Playback Controls

The playback controls in the Music app are the familiar Play/Pause, Rewind and Fast Forward buttons.

Tap the Play button to begin playing the audio.

Tap Rewind (left arrows) to move to the previous song.

Tap Fast Forward (right arrows) to move to the next song.

Tap + or − or rotate the Digital Crown to adjust the volume.

Shuffle and More

Force Touch allows you to access other options within the Music app by pressing harder on the screen. The available options vary depending on what items are on the screen at the time.

SHUFFLE, REPEAT, SOURCE, AIRPLAY

These options are visible if you Force Touch the Now Playing screen.

SOURCE, NOW PLAYING

These options are visible if you Force Touch the display while viewing lists of artists, albums or playlists.

PLAY ALL, SHUFFLE ALL, SOURCE, NOW PLAYING

These options are visible if you Force Touch the display after selecting an artist.

SHUFFLE ALL, SOURCE, NOW PLAYING

These options are visible if you Force Touch when viewing songs, or within an album or playlist.

Modifying **Music Settings**

As with other Apple Watch apps, you can customize some of the features of the Music app to suit your needs through the Apple Watch app on your iPhone.

Music in My Watch

1 Open the **Apple Watch** app on your iPhone.

2 Find **Music** in the My Watch section of the app and tap to reveal its options.

3 Configure the available options as desired.

Tap Music to see the options available.

Toggle to hide or show in Glances.

Specify which songs you want stored on Apple Watch with a playlist.

You can control how much space your music is allowed to take up on your Apple Watch.

You can also limit to a certain number of songs, regardless of file size.

APPLE WATCH MUSIC STORAGE LIMIT

Apple Watch allows you to store up to 2 GB (2 gigabytes) of audio files directly on it. Tap **Playlist Limit** to adjust the amount of your Apple Watch's built-in memory that is allocated specifically for audio file storage. To allocate a specific amount of memory, tap **Storage** and select an amount below. To allocate by a limited amount of audio files, tap **Songs** and select a number of audio files from the list provided. Allocate memory to audio files on your Apple Watch by the amount of memory or number of audio files.

Storing and Playing Audio
on Apple Watch

Apple Watch allows you to store up to 2GB of audio files directly on it. In addition, you can listen to any music stored on your iPhone when your iPhone and Watch are in range.

Storing Audio

1 Create a playlist on your iPhone to sync with Apple Watch. You can only sync one playlist at a time. Playlists can be created directly on your iPhone or using iTunes, syncing the playlist with your iPhone.

2 Connect Apple Watch to its charger.

3 Open the Apple Watch app on your iPhone. Find **Music** in the My Watch tab and tap to select it.

4 Tap **Synced Playlist** to see the playlists stored on your iPhone.

5 Tap the playlist you want to sync with Apple Watch. 'Sync Pending' will appear next to the playlist and will then change to Syncing, also displaying a percentage of the sync process.

6 Once the sync process is complete, the name of the synced playlist appears in the Music section of the Apple Watch app.

Choose a playlist to sync.

Not all songs will be synced if your playlist is longer than the Playlist Limit.

> ## Warning
> Do not use your Apple Watch during the syncing process or you'll have to start over.

Listening to Audio

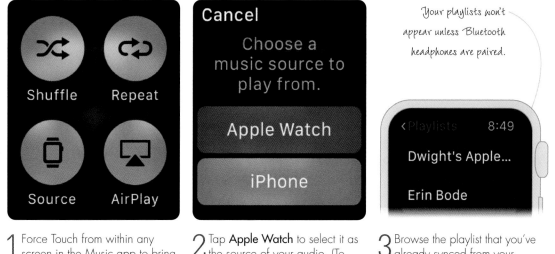

Shuffle   Repeat

Source   AirPlay

Cancel

Choose a music source to play from.

Apple Watch

iPhone

Your playlists won't appear unless Bluetooth headphones are paired.

‹ Playlists   8:49

Dwight's Apple...

Erin Bode

1 Force Touch from within any screen in the Music app to bring up a list of options. Tap **Source**.

2 Tap **Apple Watch** to select it as the source of your audio. (To browse your iPhone music, select iPhone.)

3 Browse the playlist that you've already synced from your iPhone, select an audio file and get your groove on.

WHICH DEVICE PLAYS THE AUDIO?

Whichever device is storing the audio file will be the device it is played back on when using the Music app on your Apple Watch. For example, if the source of the audio is your iPhone, the audio plays through your iPhone speakers. If the audio file is stored on your Apple Watch, then it will play the tune through your paired Bluetooth headphones.

Pairing and Unpairing
Bluetooth Headphones

In order to listen to audio stored on your Apple Watch, you'll need a pair of Bluetooth headphones. These communicate wirelessly with the Apple Watch, eliminating cumbersome cords.

Pairing Bluetooth Headphones

1 Turn on your Bluetooth headphones and make sure they are in pairing mode. (Check with the manufacturer's documentation to find out how to do so.)

2 Press the Digital Crown to get to the Home screen (if you're not already viewing it, of course). Tap the **Settings** app (the icon looks like a gear).

3 Tap **Bluetooth**. If you don't see **Bluetooth**, swipe up or down or turn the Digital Crown to scroll through the Settings options.

4 Tap the name of the Bluetooth headphones once the Apple Watch discovers them. Enter a PIN or passkey if prompted.

5 Apple Watch will notify you once the pairing is completed or if there was a problem during the process. If you run into a problem, try steps 1 to 4 again after powering off your Bluetooth headphones.

SET UP HEADPHONES QUICKLY

When you Force Touch a selection in the Music app, you are given an option to select the source of your audio, your options being the Apple Watch or your iPhone. If you don't already have Bluetooth headphones paired with your Apple Watch, you will be prompted by Music to do so. From that prompt simply tap the **Settings** button and then continue by tapping **Bluetooth** and then selecting your device.

Unpairing Bluetooth Headphones

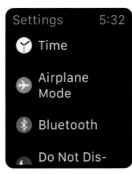

1 Press the Digital Crown to get to the Home screen (if you're not already viewing it, of course). Tap the **Settings** app (the icon looks like a gear).

2 Tap **Bluetooth**. If you don't see Bluetooth, swipe up or down or turn the Digital Crown to scroll through the Settings options.

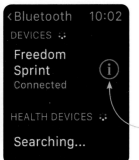

3 Tap the **Information** button to the right of the name of the Bluetooth headphones.

Tap the Information button.

4 Tap **Forget Device**. By the way, there is no further warning: the Bluetooth headphones are simply unpaired.

There is no confirmation if you choose this option.

CHAPTER 12

Weather, Maps
and Stocks

Using the **Weather App**

Weather on Apple Watch is a fine example of just how convenient your Apple Watch is to use. Just glance down at your wrist to see what the current weather is in your location.

Launching Weather

1 Press the Digital Crown once to access the Home screen. Find the **Weather** app's icon and tap it.

2 Your local weather is displayed, with the current temperature in the centre.

Accessing Weather from Glances

Many of Apple Watch's watch faces allow you to add a weather complication.

1 Open the **Clock** app, if you're not already there. Swipe up on the display of your Apple Watch to open Glances.

2 Swipe right or left to find the Weather glance and take a gander at what's going in your area. Tap the Weather glance to open the Weather app.

Checking the Weather

Current temperature

Hourly weather forecast

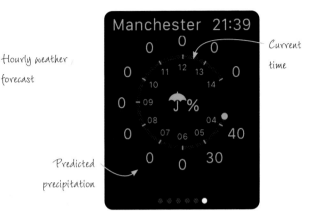

Current time

Predicted precipitation

Current Conditions
When you first open the Weather app, you are provided a look at the current conditions in your default city.

View Other Weather Information
Tap the Apple Watch display to cycle between displaying temperature, condition and chance of precipitation. You can also Force Touch the display to jump between the modes.

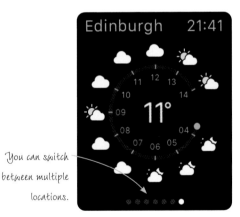

You can switch between multiple locations.

10-Day Forecast
Swipe up on the display to reveal the 10-Day Forecast for your area. Continue to swipe up to see further along in the coming days.

Multiple Locations
If you have multiple locations in your Weather app, you can swipe right or left on the display to bring them into view. Swipe down on the other locations to see their 10-Day Forecasts, as well.

Configuring Weather App
Options and Settings

The Weather app is customizable when it comes to the locations and other weather information it displays. However, Apple Watch itself isn't up to the task; you'll need your trusty iPhone from here.

Making Changes Within iPhone's Weather App

Tap the menu icon to change options.

Tap and hold a city to move it up or down.

Swipe left on a city to delete it from the list.

You have to confirm deleting cities.

The Apple Watch Weather app automatically syncs with the Weather app on your iPhone. To make changes, open **Weather** on your iPhone and tap the menu icon. All changes you make here will sync back to your Apple Watch.

Tap the scale icon to switch between Celsius and Fahrenheit.

Tap the + icon to add a city.

Making Changes Within iPhone's Apple Watch App

1 Open the **Apple Watch** app on your iPhone and go to the **My Watch** tab. Scroll to the **Weather** icon and tap it.

2 Toggle **Show in Glances** On (green) to show the Weather in your Apple Watch's Glances, or to Off (black) to disable the feature.

3 To change the default city, tap the **Default City** option and select the name of the location you want to be your default. A tick will appear to the right of it.

THIRD-PARTY WEATHER APPS

The Apple Watch Weather app does a very good job with giving you the latest information, as it's powered by The Weather Channel. However, there are other excellent third-party apps available as well that have components for both iPhone and Apple Watch, including:

- Yahoo! Weather!
- Dark Sky
- Weather Underground
- The Weather Channel
- AccuWeather

Search for these and other weather-related apps in the Apple Watch App Store and try them out for yourself.

Navigating the **Maps** App

The Maps app on Apple Watch makes it easy to get accurate directions to your destination, and your Watch will direct you through every turn.

Maps Basics

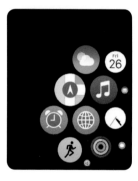

To get started with Maps, press the Digital Crown once to access the Home screen (if you're not already there). Find the **Maps** icon and tap to open it. The first thing you're presented with is a map of your current location.

The city that you're in is indicated at the top left of the display.

Drag the map around with your finger to have a look at your surroundings.

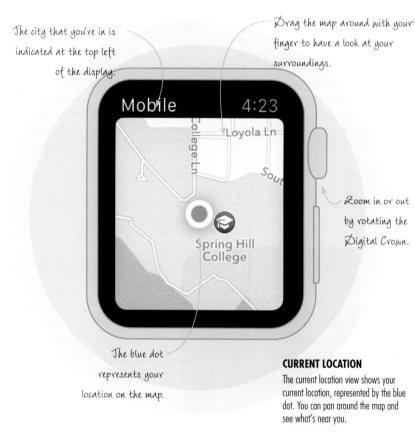

Zoom in or out by rotating the Digital Crown.

The blue dot represents your location on the map.

CURRENT LOCATION
The current location view shows your current location, represented by the blue dot. You can pan around the map and see what's near you.

LOCATION INFORMATION

Tap locations or pins to see more information about them. Rotate the Digital Crown to read more. Tap the left-pointing arrow in the upper left corner of the display to return to the map.

Tap a location to get more information about it, like hours or phone numbers.

Tap the Current Location arrow to jump back to your current location.

DROPPING A PIN

Drop a pin if you want to mark a location or get more information about it. Touch and hold a place on the map until the pin drops on it. Then tap the pin to find more information about the location. This is useful in finding addresses of locations.

To delete a pin, scroll down on the information screen and tap Remove.

Tap and hold to drop a pin.

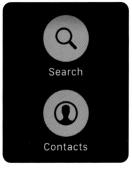

CONTACT LOCATIONS

Force Touch the Maps display and tap Contacts to search your contacts list for people and destinations. Just tap their address to get directions.

THE MAPS GLANCE

Want to get a quick look at your current location and what else is around? Just swipe up on your Apple Watch display from the Clock app to open Glances, and then swipe right or left until you find the Maps glance.

Searching for Locations and Getting Directions

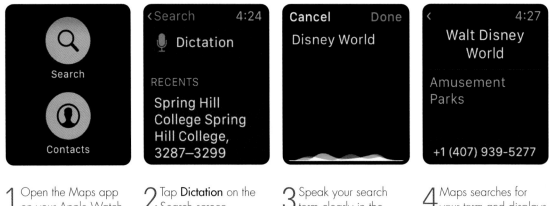

1 Open the Maps app on your Apple Watch. Force Touch the display to see options and tap **Search**. (You can also get directions to a contact's address.)

2 Tap **Dictation** on the Search screen.

3 Speak your search term clearly in the direction of your Apple Watch's microphone and tap **Done**.

4 Maps searches for your term and displays the results. Scroll down to see contact information for your search term, directions options and a map of the term's location.

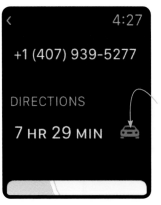

Your estimated arrival time will be shown here.

Only driving directions are shown for long distances. Shorter distances may offer other transit options.

5 Tap the directions option you desire and Maps will plot a course from your current location to your destination.

6 Tap **Start** to begin your trip or **Clear** to wipe out the results and start over.

Taking Direction from Maps

Arrival Time
After starting directions, you'll see an ETA (estimated time of arrival) in the upper left corner. Tap the ETA to also see an estimate of how much time it will take to reach your destination. Tap again to go back to the ETA.

Force Touch to Stop
Should you want to quit the directions at any point, just Force Touch the display and tap the Stop Directions button. You can also tap the Call Destination button to ring the people you're going to see.

Turn-by-Turn Directions
Maps provides you with turn-by-turn directions, guiding you every step of the way. Your directions are broken down into slides based on where you are to make turns.

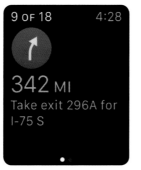

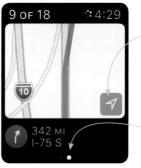

You can see which slide number you are viewing in the top left of the display.

View Directions or Map
As you go on your way, Maps will move from slide to slide in the directions. You can also rotate the Digital Crown or swipe up to scroll backwards or forwards in your direction slides. To view a Map, swipe left.

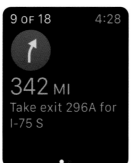

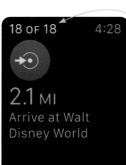

Tap to return to your current point in the route.

You can switch between the directions and map by swiping left or right.

DON'T IGNORE THOSE TAPS!

As you progress through your journey, Apple Watch will alert you to turns and your arrival through a series of taps on your wrist. You will receive 12 steady taps when it's time to turn right at the next intersection, 3 sets of 2 taps each when you need to go left, and you're given a steady vibration upon arrival at your destination.

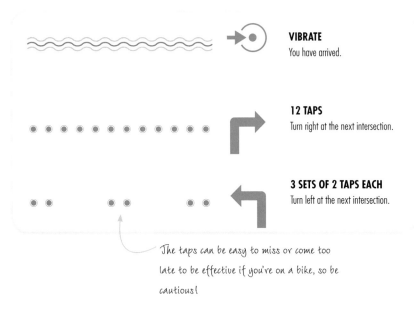

VIBRATE
You have arrived.

12 TAPS
Turn right at the next intersection.

3 SETS OF 2 TAPS EACH
Turn left at the next intersection.

The taps can be easy to miss or come too late to be effective if you're on a bike, so be cautious!

Slide up on the Map icon to continue on.

USING HANDOFF WITH MAPS

As you know, Apple Watch and iPhone are partners in assisting you. They are quite adept at getting you to your destination with their respective Maps apps, but did you know that those apps can work together for your benefit? The devices utilize Handoff to move information from one to the other, and back again if you please.

1 Open **Maps** on your Apple Watch and find a location.

2 Wake your iPhone (don't unlock it) and you'll find the Maps icon in the lower left corner of the screen.

3 Swipe up on the Maps icon on your iPhone to open its Maps app to the same location.

You can perform this same kind of Handoff whether you're getting directions, finding locations or just browsing around.

Setting Maps Options

There are a couple of options that you can set for the Maps app on Apple Watch, but you have to configure them on your iPhone.

1 Open the **Apple Watch** app on your iPhone and go to the **My Watch** tab.

2 Find the **Maps** icon in the list of options and tap to open it.

Toggle On to see the Maps glance when you swipe up on the display when using the Clock app. Toggle to Off to disable the glance.

Tap the Maps icon to set options.

Toggle On to be alerted on your Apple Watch when it's time to make a turn. Toggle to Off to not receive alerts.

Ask Siri for Directions

You can always engage Siri, and ask her help in getting directions. It's so easy: simply ask Siri what you want to know. For example, say 'Hey Siri, directions to Disney World,' and she'll happily provide them for you.

The **Stocks** App

The Apple Watch is happy to give you the latest market prices in a clear and concise manner with just a glance at your wrist.

Viewing Stocks

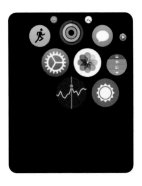

1 Open the **Stocks** app by tapping its icon on the Home screen of your Apple Watch.

2 Swipe up and down on your Apple Watch's display to see your list of stocks.

3 Tap a stock to see detailed information such as prices, market cap and more.

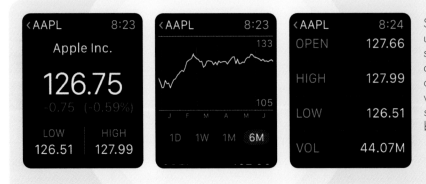

Swipe or scroll down using the Digital Crown see a graph displaying a stock's performance over periods of a day, a week, a month or up to six months (tap the icons below the graph).

Setting Options for Stocks

1 Open the **Apple Watch** app on your iPhone and go to the **My Watch** tab. Scroll to the Stocks icon and tap it.

2 Toggle the **Show in Glances** switch to On (green) to enable the Stocks glance. Turn off the glance by toggling the switch to Off (black).

3 Tap the **Default Stock** option to select a stock as your default. This means it will be the stock that is shown in the Stocks glance and in the Stocks complications on some watch faces.

4 Choose which information to show about your default stock on the watch face complications by tapping your selection.

Adding and Removing Stocks

To make changes to the stocks displayed in the Stock app on Apple Watch, you'll need to go through the Stock app on your iPhone. All changes you make here will sync back to your Apple Watch.

Tap the + button to add a stock.

Move stocks up and down with the grabber.

Tap the red circle to delete a stock.

You have to confirm deleting stocks.

Tap the menu icon to change options.

CHAPTER 13

Remote Control Features

Using Remote with **Apple TV**

You can use the Remote app on Apple Watch to control Apple TV, making it easy to browse networks and select programmes without digging behind the sofa cushions for the remote control.

Set Up Remote to Control Apple TV

1 Open the **Remote** app on your Apple Watch by tapping its icon on the Home screen.

2 Tap **Add Device** and a passcode will appear on the display.

3 On your Apple TV, go to **Settings** > **General** > **Remotes**.

4 Select your Apple Watch in the list of devices and enter the passcode that is shown on your Apple Watch.

Use the passcode shown on your Remote app.

How to Control Apple TV

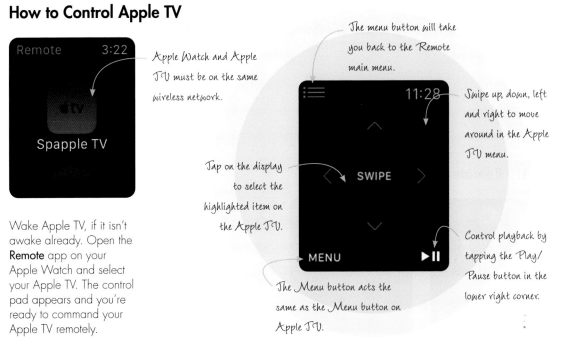

Apple Watch and Apple TV must be on the same wireless network.

The menu button will take you back to the Remote main menu.

Swipe up, down, left and right to move around in the Apple TV menu.

Tap on the display to select the highlighted item on the Apple TV.

The Menu button acts the same as the Menu button on Apple TV.

Control playback by tapping the Play/Pause button in the lower right corner.

Wake Apple TV, if it isn't awake already. Open the **Remote** app on your Apple Watch and select your Apple TV. The control pad appears and you're ready to command your Apple TV remotely.

HOW TO UNPAIR APPLE WATCH AND APPLE TV

1 Go to **Settings** > **General** > **Remotes** on your Apple TV.

2 Select **Apple Watch** in the iOS Remotes section and then select Remove on the next screen.

3 Open the **Remote** app on your Apple Watch and tap the **Remove** button when you see the lost connection message.

Using Remote with **iTunes**

iTunes is Apple's hub for all your iOS devices and your gateway to music, films, podcasts and much more. Now you can use the Remote app on Apple Watch to control iTunes right from your wrist.

Set Up Remote to Control iTunes

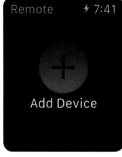

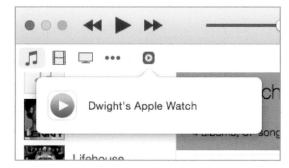

1 Open iTunes on your Mac or PC.Open the **Remote** app on your Apple Watch by tapping its icon on the Home screen.

2 Tap **Add Device** and a passcode will appear on the Apple Watch display.

3 A remote icon will appear in the tab bar of iTunes 12 or newer (upper left corner of the window); click the remote icon. If you're using iTunes 11 or prior, the remote icon will appear in the upper right under Search.

4 Enter the passcode from your Apple Watch into the fields provided in iTunes. iTunes will provide a verification screen once the Apple Watch is connected.

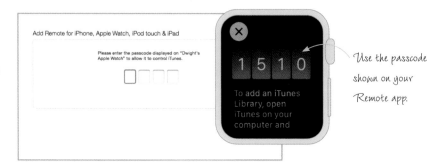

Use the passcode shown on your Remote app.

How to Control iTunes

Apple Watch must be on the same wireless network as the computer with iTunes.

Tap the menu button to go back to the Remote menu.

You can skip, play and pause just like in the Music app.

You can also control the volume.

Open the **Remote** app on Apple Watch. Tap the iTunes library you want to control. Use the controls on the display to control playback of audio in iTunes.

HOW TO REMOVE AN ITUNES LIBRARY FROM REMOTE APP

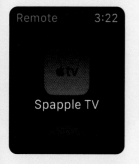

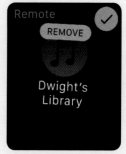

1 Open the **Remote** app on Apple Watch.

2 At the list of devices, Force Touch the Apple Watch display to access the Edit button. Tap the **Edit** button.

3 Items in the device list will begin wiggling. Tap the **x** on the library you want to remove.

4 Tap **Remove** on the library to give it a quick exit from your devices list.

CHAPTER 14

The Apple Watch App Store

Accessing the Apple Watch
App Store

The Apple Watch App Store is the place to go to find third-party apps that are designed to work on your Apple Watch. You can access the Apple Watch App Store from your iPhone.

Apps for Your Watch

Apps for the Apple Watch are typically extensions of apps that you use on your iPhone. That doesn't always mean that their functionality is limited (although that may sometimes be the case); in fact, some apps are designed for use mainly on your Apple Watch. You'll find apps in the Apple Watch App Store to satisfy your work and play needs in the following categories:

- Business
- Finance
- Food & Drink
- Games
- Health & Fitness
- Lifestyle
- News
- Productivity
- Social Networking
- Travel
- Utilities

New apps are being added on a seemingly daily basis, so you'll want to check the Apple Watch App Store frequently for the latest and greatest apps.

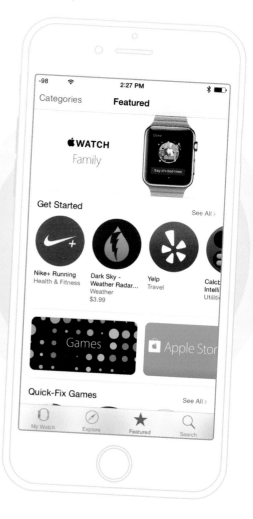

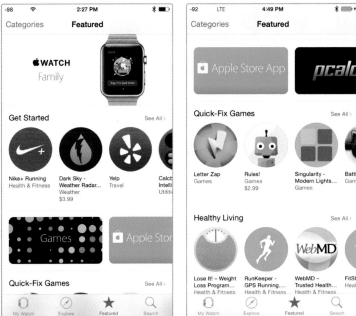

1 Open the **Apple Watch** app on your iPhone. Tap the **Featured** tab at the bottom of the screen.

2 Swipe up and down to see apps that Apple features in the Apple Watch App Store. Swipe left and right on a category to see more featured apps.

3 Tap **See All** to the right of the screen when viewing a category to see all the category's featured apps in a convenient list.

The App Store indicates iPhone apps that offer Apple Watch apps.

FIND APPLE WATCH APPS USING THE IPHONE APP STORE

You can use the iPhone App Store to find apps that also work with Apple Watch, should you be so inclined. When you come across an app that also works with the Apple Watch, you'll see 'Offers Apple Watch App' under the icon of the app.

Navigating **the App Store**

You can find apps for Apple Watch in the Apple Watch App store, which is accessed through the Apple Watch app on your iPhone. Browse by category or search for specific apps.

Browse Apps by Categories

1 Open the **Apple Watch** app on your iPhone.

2 Tap **Featured** at the bottom of the My Watch tab.

3 Tap **Categories** in the upper left of the screen.

4 Tap the category in the list that suits your needs to open its screen.

5 Browse the apps in the category selected. Swipe up and down to scroll to subcategories, and swipe left or right to browse them.

Tap See All to the right of a subcategory to see all the apps it contains in a convenient list.

Search for Apps

1 Open the **Apple Watch** app on your iPhone.

2 Tap **Search** at the bottom of the My Watch screen.

3 Tap the Search field at the top of the screen to open the onscreen keyboard.

If you don't know the exact name, you can use keywords like 'note', too!

4 Type your search terms into the search field. The screen will populate with the best matches to your search criteria as you type. Once you see what you're searching for in the list, tap it.

5 Tap the name of the app to see more information about it, or tap the price of the app (the price will say 'Get' if it's free or 'Open' if you already have it installed on your iPhone).

6 Tap **Install** to install the app on your iPhone. You may be prompted to put in your Apple ID and password.

Installing and Removing Apps

You'll probably find it easiest to manage the installation and removal of apps for your Apple Watch through the Apple Watch app on your iPhone.

Download Apps on iPhone

The first step to getting an app onto your Apple Watch is downloading the app on your iPhone. Find the app you want to install using the Apple Watch App Store or the iPhone App Store, and download it. Once the download is complete, you should see the app icon appear on both your iPhone and the Home screen of your Apple Watch. If the app icon does not appear on the Home screen of your Apple Watch, you may need to manually install it by enabling the app through the My Watch tab of the Apple Watch app for iPhone.

Tap the button to download the app.

Install and Remove Apple Watch Apps on iPhone

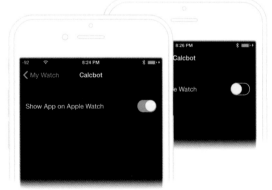

Open the **Apple Watch** app on your iPhone and go to the **My Watch** tab. Scroll down the list of options until you see the name of the app you installed on your iPhone, and then tap the icon. To access the app on your Apple Watch, toggle the **Show App on Apple Watch** switch to On (green). To remove the app from Apple Watch, toggle the switch to Off (black). This does not remove the app from your iPhone; you can add it back to your Apple Watch later if desired.

Remove Apple Watch Apps Directly

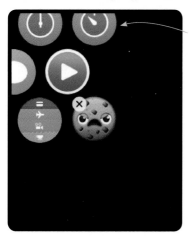

First-party apps from Apple cannot be removed from Apple Watch.

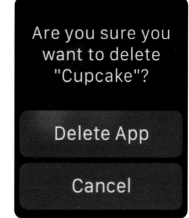

1 Tap and hold the icon of the app you want to remove. When the app icons start wiggling, tap the **x** on the icon you want to delete.

2 Confirm deletion of the app when prompted and the app will be removed from your Apple Watch. The app will remain on your iPhone and you can add it back to your Apple Watch at another time through the Apple Watch app.

ADJUST APP SETTINGS
Many apps allow minor adjustments so that you can customize how they work with your Apple Watch.

1 Open the **Apple Watch** app on your iPhone and go to the **My Watch** tab.

2 Scroll down the list of options until you see the name of the app you wish to adjust, and then tap its icon.

3 Make any adjustments you deem necessary. Changes are immediately applied to the app on Apple Watch.

Third-party apps can be shown in Glances, and may even have notification options.

CHAPTER 15

Troubleshooting

Common
Troubleshooting Measures

When it comes to troubleshooting problems with electronic devices, there are three simple actions that will often correct the problem: restarting, resetting to factory settings and restoring settings.

Restart Apple Watch

PRESS AND HOLD

Normal Restart
Restarting your Apple Watch is one of the most effective and useful troubleshooting measures.

1 Press and hold the side button until the sliders appear on the display.

2 Slide the **Power Off** slider all the way to the right. Your Apple Watch will turn off.

3 Press and hold the side button until the Apple logo appears and then wait for the Apple Watch to power up.

Once you see the Apple logo, release the buttons.

PRESS AND HOLD SIMULTANEOUSLY

Force Restart
If you aren't able to normally restart your Apple Watch, you can force it to restart.

1 Press and hold both the side button and the Digital Crown for at least 10 seconds.

2 Release both buttons simultaneously once the Apple logo appears.

Reset Apple Watch to Factory Settings from iPhone

1 Open the **Apple Watch** app. Select the My Watch tab at the bottom of the screen. Find **General** in the list of options and tap it. Scroll down and tap **Reset** at the bottom of the list.

2 Tap **Erase All Content and Settings** and confirm your selection by tapping **Erase All Content and Settings** again when prompted. You could also end the carnage by tapping **Cancel**.

Unpair Before Resetting

When you reset your Apple Watch, you wipe out all of your personal data and settings. If you want to preserve those settings, unpair your Apple Watch from your iPhone before resetting it. This causes your Apple Watch to make one last backup with your iPhone.

Reset to Factory Settings from Apple Watch

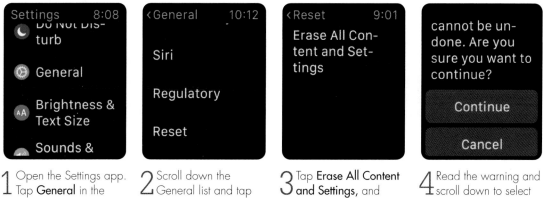

1 Open the Settings app. Tap **General** in the Settings app list of options.

2 Scroll down the General list and tap **Reset**.

3 Tap **Erase All Content and Settings,** and enter your passcode if prompted to do so.

4 Read the warning and scroll down to select **Continue** to go ahead with the reset or **Cancel** to put the brakes on.

Restore Apple Watch Settings from a Backup

If you've had to start all over with your Apple Watch you have two choices: continue setting up the Watch as if it were brand new or restore your Apple Watch to its former personalized glory by restoring it with a backup. Apple Watch backs itself up to your iPhone automatically (when they are in range of one another,

of course), but it's a good idea to unpair and re-pair the devices from time to time. Why? Because unpairing your Apple Watch and iPhone creates a backup, and you know the latest information is being saved.

1 Power on your Apple Watch.

2 Open the **Apple Watch** app on your iPhone.

You'll have to go through the whole pairing process before restoring from backup.

3 Tap the **Start Pairing** button and follow the instructions to pair your Apple Watch with your iPhone.

4 Once pairing is complete, you will be asked to Set Up as New Apple Watch or Restore from Backup: tap **Restore from Backup**.

You may have multiple backups.

5 Tap the backup you want from the list provided. The most recent backup is at the top of the list.

6 Agree to the Terms and Conditions, enter your Apple ID and continue with the rest of the set-up process until it is completed. When you see the Apple Watch Is Ready screen, you'll know it's done and your settings have been restored to your Apple Watch.

WHAT INFORMATION IS SAVED DURING A BACKUP?
A backup will include quite a lot of your Apple Watch's information, but not everything. Backups do include:

- General settings (your watch face preference, brightness and sound settings, and the like)
- Your time zone
- Your default system language
- Settings you made for Calendar, Stocks, Weather and Mail
- Health & Fitness information
- Information and settings you've configured for apps

Backups do not include credit or debit cards used for Apple Pay, passcodes, synced playlists or Workout and Activity calibration information.

Forgotten Passcode

 If you've forgotten your passcode, all you can do is erase your Apple Watch content. Once erased, you can restore your settings from a backup and set a new passcode.

Erase Content Using iPhone

1 Open the **Apple Watch** app and go to **My Watch**.

2 Find and tap **General,** and then find and tap **Reset**.

3 Tap **Erase All Content and Settings,** and then tap again to confirm.

You'll have to go through the pairing process again.

Erase Content Using Apple Watch

Force Touch until you get the option to erase.

1 Connect your Apple Watch to its charger. Press and hold the side button until the power sliders appear.

2 Force Touch the Power Off slider until you see the red **Erase all contents and settings** button, and then tap it to erase your Apple Watch.

Problems with **Apps**

If an app you've installed is missing, failing to launch correctly or behaving erratically, you can usually solve the problem by restarting or reinstalling the app.

Restart

Press and hold the side button until you see the sliders, then slide the Power Off slider to the right. After the Watch goes off, press and hold the side button until the Apple logo appears. Try launching the problematic app again.

Remove and Reinstall App

You can also remove an app by pressing and holding it until the icons start to wiggle. Then tap the x to remove the app.

1 Open the **Apple Watch** app on your iPhone and go to the **My Watch** tab.

2 Scroll down to the list of installed apps and tap the offending one.

3 If the **Show App on Apple Watch** toggle switch isn't On (green), tap it to begin installing the Apple Watch version of the app on your Watch. The app should now appear on your Apple Watch.

4 If the **Show App on Apple Watch** toggle switch is On (green), tap it to uninstall the app from your Apple Watch. Toggle the switch back to On to reinstall the app and try again.

Apps Worked Fine Before a WatchOS update

Sometimes apps may misbehave after you've performed an update to your Apple Watch's operating system. If this is the case, check the App Store to see if the app has been updated for the newest WatchOS version and install the updated app to see if the issue is corrected.

Apple Watch **Won't Turn On** or Respond

If your Apple Watch isn't responding to your pokes and touches, or simply won't turn on, don't panic. There are a few reasons why this may happen, and it's likely it can be resolved.

Reasons for Unresponsiveness

- The battery is dead – connect it to its charger.
- It's in Power Reserve mode.
- Screen Curtain is enabled.
- The screen is frozen.
- There is a serious system problem (very highly unlikely, but still a possibility).

Power Reserve Mode

If your Apple Watch won't wake when you tap it or when you flick your wrist, it may just be in Power Reserve mode. In that case:

- If all you see is the time displayed, press and hold the side button until you see the Apple logo to exit Power Reserve mode.
- If you see a small red lighting bolt to the left of the time on your Apple Watch's display, the Watch needs to be charged. Connect the Watch to its charger and all will be right with the world.

1:30

If you only see the time, Power Reserve is turned on.

Screen Curtain Might be Enabled

If your display is totally black but you hear Apple Watch describing elements that are normally on your screen when you tap the display, you may have Screen Curtain enabled. To resolve:

1 Open **Apple Watch** app on your iPhone and go to **My Watch**.

2 Go to **General** and tap **Accessibility**, then tap **VoiceOver**.

3 Toggle the switch for **Screen Curtain** to Off (black) and do the same for VoiceOver.

If Screen Curtain is enabled, you'll hear Siri describing elements on the screen, but the screen stays off.

Frozen or Unresponsive Display

Should the screen appear to be frozen (unresponsive to anything), try forcing it to restart.

1 Press and hold both the side button and the Digital Crown for at least 10 seconds.

2 Release both buttons simultaneously once the Apple logo appears.

PRESS AND HOLD SIMULTANEOUSLY

When All Else Fails

If a forced restart doesn't correct the problem, you will want to contact Apple technical support directly.

Cannot Connect to My iPhone

As you know, iPhone is essential to your Apple Watch experience, so if the two aren't talking, then your world is a bit discombobulated. Bluetooth just might be the issue.

Bluetooth is not connected.

Check that Bluetooth is turned on.

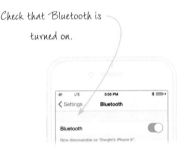

You can tell if the connection is broken between the devices if a red iPhone icon appears at the top of your Apple Watch display. There are two ways to check your iPhone's Bluetooth setting (you can't disable Bluetooth on your Apple Watch).

Bluetooth in Control Centre
Swipe up from the bottom of your iPhone's screen to open the Control Centre. If the Bluetooth icon is white, Bluetooth is enabled. If the icon is grey, it's disabled. Simply tap the grey Bluetooth icon to enable it.

Bluetooth in Settings
Open the Settings app, tap Bluetooth in the list and toggle the Bluetooth switch to On (green).

IF YOU'RE STILL HAVING ISSUES

1 **Restart your Apple Watch.** Press and hold the side button until you see the sliders, then slide the Power Off slider to the right. After the Watch goes off, press and hold the side button until the Apple logo appears.

2 **Force a restart of your Apple Watch.** Press and hold the Digital Crown and the side button simultaneously until you see the Apple logo on the display.

3 **Restart your iPhone.** Press and hold the side button until you see the Power Off slider. After your phone goes off, press and hold the side button until the Apple Logo appears.

4 **Unpair and re-pair your Apple Watch and iPhone.** Unpair your Apple Watch by going through the My Watch tab of the Apple Watch app, then re-pair it with your iPhone.

Apple Watch Is **Not Charging**

Your Apple Watch depends on a battery, so if that battery can't get a charge for some reason your Apple Watch is going to be very sad (and so are you).

Restart Apple Watch

1 Press and hold the side button until the sliders appear.

2 Drag the **Power Off** slider all the way to the right.

3 Turn your Watch back on by pressing and holding the side button until the Apple logo shows up on the display.

4 Try charging your Apple Watch again.

Force Apple Watch to Restart

1 Press and hold the Digital Crown and the side button together until the Apple logo appears.

2 Try charging your Apple Watch again.

Reset Apple Watch

Use the steps listed earlier in this chapter, and then attempt to charge it. At this point, if you're still having difficulty charging, there could be several reasons:

- A faulty charger
- A faulty battery
- Another hardware defect

Contact Apple technical support as soon as possible to rectify the situation.

The Digital Crown isn't Responding Properly

Apple Watch's Digital Crown is really an amazing piece of technology, but it's not immune to problems. If your Digital Crown isn't responding as it should, these measures may help.

Manipulate the Digital Crown

First, try spinning and pressing the Digital Crown multiple times to free it from any dirt or other foreign body that may be impeding it.

Don't use compressed air to clear the Digital Crown; it could cause damage.

Rotate the Digital Crown both ways, to free it of debris.

Restart Your Apple Watch

Normal Restart

1 Press and hold the side button until the sliders appear on the display.

2 Slide the **Power Off** slider all the way to the right. Your Apple Watch will turn off.

3 Press and hold the side button until the Apple logo appears and then wait for the Apple Watch to power up.

PRESS AND HOLD

Force Restart

Try a forced restart, if possible. You may not be able to force restart your Apple Watch if the Digital Crown is not responding.

1 Press and hold both the side button and the Digital Crown for at least 10 seconds.

2 Release both buttons simultaneously once the Apple logo appears.

PRESS AND HOLD SIMULTANEOUSLY

Clean the Digital Crown

Press the band release button on the back of the watch.

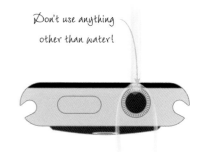

Don't use anything other than water!

1 Turn off your Apple Watch and disconnect it from the charger if it's connected.

2 Remove the bands from the Apple Watch, especially if they're leather.

3 Run some warm water from a faucet (lightly, of course) over the Digital Crown. Have no fear: Apple themselves came up with this fix.

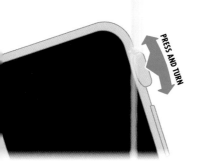

Allow time to air dry, especially in the channels for the bands.

4 Rotate and press the Digital Crown while holding it under the water for about 10 to 15 seconds.

5 Remove excess water with a soft cloth and allow to completely air dry. (Do not use a hair dryer or other device to speed drying.)

6 Turn on your Apple Watch and test to see if it responds correctly to use.

Response Time is Slow

Computer users know full well that sometimes things can get a little slow. Your Apple Watch can slow down, too. If this happens, an app may be the culprit.

Press and hold the app until it begins wiggling, and tap the x to remove.

Reasons for Slowness

- Poorly-developed apps can quickly suck the life from your Apple Watch. These apps can may be slow themselves, or they may steal resources from other apps.
- There may be an electronic glitch that is causing your Apple Watch to run slowly.
- A third-party app that requires intensive processing may run a little slower than you'd like. This is due to the fact that they must use the iPhone to handle their computing processes, and if there's a lot going on, the lag time may simply be unavoidable.

Things to Attempt

Uninstall and Reinstall the App
If you suspect an app is causing the issue itself, try uninstalling and reinstalling it, or simply get rid of it altogether and find a better alternative.

Restart Your Apple Watch
Restart your Apple Watch normally. If it is frozen, force a restart of your Apple Watch by pressing and holding the Digital Crown and side button simultaneously until the Apple logo appears.

Reset Your Apple Watch
If it's still slow after removing the app and restarting, you may need to reset your Apple Watch to factory settings.

Additional **Help**

If you find yourself in need of additional help, whether with your Apple Watch or an app, there are a few places you can go to to connect with other Apple Watch users and specialists.

Apple Resources

- Check out the Apple Watch support page: apple.com/uk/support/watch.
- Go to the discussions forum to find other Apple Watch wearers who may have experienced and resolved issues similar to yours: discussions.apple.com/community/watch.

Contact Apple Support

1 Go to the Apple Watch Contact Support page: apple.com/uk/support/watch/contact.

2 Click the Get Started link under the Apple Support section, or type this into your browser's URL field: getsupport.apple.com/GetproductgroupList.action (click the Apple Watch link once you get there).

3 Select a support topic from the list and get the help you need from the knowledgeable and friendly staff at Apple.

Don't Forget Your Local Apple Store

If you're near an Apple Store and are experiencing issues with your Apple Watch, just pop into the store and get help from one of the specialists who are there just to help you and other loyal Apple customers.

Glossary

Accessibility Shortcut A setting that allows you to triple click the Digital Crown to quickly enable Apple Watch's accessibility features: VoiceOver and Zoom.

Airplane Mode A setting that disables wireless communications for your Apple Watch, but leaves other functions enabled.

apps Applications that are designed to be used on Apple Watch and iPhone.

Apple ID The personal account you set up with Apple that allows you to purchase or rent items in the Apple Store, and also keeps content synchronized across your devices.

Apple Pay A system of in-store payment that allows you to store credit or debit cards on your Apple Watch or iPhone so that you can simply hold them up to a special scanner to complete purchases.

Apple Store An online store that sells Apple devices, software and accessories, as well as those from many Apple-friendly third-parties.

Apple Watch app An app specifically designed for setting up and making configuration changes to your Apple Watch from your iPhone. You can also use it to update Apple Watch's Watch OS, find and install third-party apps and more.

band release buttons Buttons on the underside of the Apple Watch that you push to release the bands currently installed. They allow for quick and simple band installation and removal.

Bluetooth Wireless technology employed by Apple to pair the Apple Watch with an iPhone.

complications Features on a timepiece other than those for simply telling the time. For example, alarms, moon phases, chronographs, etc.

Do Not Disturb A mode in which alerts and calls won't be displayed on Apple Watch; however, alarms will still notify the wearer.

Digital Crown The dial on the side of the Apple Watch. Provides simple navigation by turning, or press it to return to the Home screen and perform other functions.

Digital Touch Communication technology new to Apple Watch that allows you to send custom animations and taps to other Apple Watch users.

faces The design and layout of the timekeeping components on the Apple Watch. Apple Watch includes several faces, and most are easily and extensively customizable, whereas traditional watches only have one face.

Force Touch Technology in Apple Watch that allows it to sense the pressure of a touch on the display, greatly extending the range of navigational techniques available to the wearer.

friends A list of the people that you contact most frequently. Access your friends list by tapping the side button once, scroll through them using the Digital Crown, and communicate through phone call, text or Digital Touch.

Glances Simple summaries of information that you can easily access by swiping up on the Apple Watch face. Scroll through available Glances by swiping to the left or right. Glances can provide quick flashes of information such as weather, calendar items, stocks and more.

Handoff The ability to start an activity on your Apple Watch and seamlessly transfer it to your iPhone. Handoff works with emails, phone calls, text messages and more.

haptics Any form of communication that involves touch. For the Apple Watch, haptics include such technologies as Digital Touch and the Taptic Engine.

heart rate sensor Located on the underside of Apple Watch, this sensor relays the wearer's heart rate information to health and fitness apps, such as Workout.

Home screen The main screen of the Apple Watch, which displays icons for apps you've installed.

iCloud Apple's cloud-based server that acts as a backup and synchronization tool for your digital content, providing access to your content on any device you are signed in to. For example, you can store pictures in iCloud and access them on your iPhone, Apple Watch, iPad, Mac or on any computer with a modern web browser.

notifications Items that appear on your Apple Watch display to notify you of events as they happen. Events include new emails, text messages, reminders, invitations and more. View stored notifications by swiping down on the display.

passcode A numerical code that you can set to prevent unauthorized users from accessing your Apple Watch.

Power Reserve mode The mode that disables all functions on the Apple Watch other than keeping the time. Apple Watch will automatically switch to this mode when battery life drops at or below ten percent.

side button The narrow button on the side of the Apple Watch. Tap once or multiple times to access various features of Apple Watch.

Siri The virtual assistant developed by Apple. Siri allows you to speak commands or ask questions using your Apple Watch by prefacing them with 'Hey, Siri' followed by your query. For example, 'Hey, Siri, what is the weather like right now?' Siri will instantly bring up the current weather conditions on the Apple Watch's display.

storage The amount of memory on your Apple Watch that can be used to store data, such as music or photos.

Taptic Engine Technology within Apple Watch that provides the wearer with taps to alert them to certain events or notifications.

VoiceOver An accessibility feature that allows you to navigate Apple Watch's interface with spoken help, even if you're unable to see the display. Also enables a special way to navigate the interface by tapping and double tapping icons.

Watch OS The operating system developed by Apple to power Apple Watch.

World Clock A feature that allows you to see what time it is in any location in the world.

Wrist Raise Technology built into Apple Watch that causes the display to turn on when you raise your wrist (in the same motion that one would use to glance at the time on a standard watch).

Zoom An accessibility feature that magnifies items on the display, making them easier to see. When Zoom is enabled the wearer can double tap with two fingers on the display to zoom in on items, and then use the Digital Crown to move around the screen.

Additional **Resources**

If you find yourself in need of information beyond the scope of this book, there are lots of other great resources that you can check out. The following are some recommended Apple Watch resources.

Apple.com

apple.com/uk/watch

Apple's own Apple Watch web page is a great place to check out Apple Watch videos and explore the different models. The Apple Watch Store is the place to go for not only ordering your Apple Watch, but also to purchase replacement bands and other accessories for your timepiece.

Apple Support

apple.com/uk/support/watch

Apple's support page for Apple Watch is where you want to go to find out tips and tricks for using your Apple Watch. To download a PDF of the Apple Watch User Guide, enter the following URL into your browser: manuals.info.apple.com/en_US/apple_watch_user_guide.pdf

Apple Watch Discussions Forum

discussions.apple.com/community/watch

Yet another Apple site is the Apple Discussions Forum for Apple Watch, where other Apple Watch enthusiasts collaborate on their experiences with the coolest device in tech. There are three distinct sections in the Apple Watch communities: Using Apple Watch, Apple Watch Hardware and Apple Watch Accessories.

Apple Pay Discussion Forums

discussions.apple.com/community/apple_pay

There is also a discussions community for Apple Pay. The three areas in its community are: Setting Up Apple Pay, Using Apple Pay in Stores and Using Apple Pay within Apps.

Apple ID Support

apple.com/uk/support/appleid

Should you have any issues or questions about your Apple ID, this is the site to find the answers.

WatchAware

watchaware.com

WatchAware is a great site that is only concerned with offering its readers with truly useful information on the Apple Watch. WatchAware focuses on three primary aspects of the Apple Watch universe: latest news, editorials and apps. These folks know what you, as an Apple Watch aficionado, really want to read about, and they stick to the plan.

Mix Your Watch

mixyourwatch.com

This website's only reason for existence is to allow you to do what Apple doesn't: mix and match the various bands Apple offers with each of the different Apple Watch models. You can get an instant preview of what your particular model will look like with the band of your choice. This site is just plain old fun, to be honest, and it removes the guesswork of buying a new band.

iMore

imore.com/apple-watch

iMore is one of the top Apple news and reviews sites on the web, and its coverage of Apple Watch is just as reliable as it has been for other Apple products.

Mac Life

maclife.com/tags/apple_watch

Mac Life is another news and reviews site, and it offers complete coverage of all things Apple, including Apple Watch.

Macworld

macworld.com

Macworld has been at the forefront of Apple news and events for decades, and it is on board the Apple Watch bandwagon as well. The experts at Macworld offer the best tips and reviews for Apple users worldwide. Although the good people at Macworld are certainly Apple fans, you can trust these folks to offer honest advice and information about Apple products and related third-party offerings.

Index

About the Author

Dwight Spivey has been a self-described Apple expert for nearly 20 years. He's the author of *How to Do Everything: Pages, Keynote & Numbers for OS X and iOS, How to Do Everything: Mac® OS X® Mountain Lion, iPhone™ & iPod® touch QuickSteps, Macs Translated For PC Users* and many more books on the Mac, iPad, iPhone and Microsoft Office. His technology experience is extensive, consisting of Mac, Linux, and Windows operating systems in general; Apple and Android devices; desktop publishing software; laser printers and drivers; colour and colour management and networking.